FORTNITE

THE **ESSENTIAL** GUIDE TO **BATTLE ROYALE**
AND OTHER SURVIVAL GAMES

This book is book is available in quantity at special discounts for your group or organization.
For further information, contact:

Triumph Books LLC
814 North Franklin Street
Chicago, Illinois 60610
Phone: (312) 337-0747
www.triumphbooks.com

Printed in U.S.A.
ISBN: 978-1-62937-679-0

Content packaged by Mojo Media, Inc.
Joe Funk: Editor
Samantha M Skinner: Writer
Jason Hinman: Creative Director
Jack Hinman & David Woodburn: Gaming Consultants

CONTENTS

STARTING OUT

Fortnite: Battle Royale came in with a BANG and immediately began offering the same sort of gameplay that PUBG was already famous for. It's important to note that Fortnite does also have some big differences that really help it stand out, and that have actually made the game more successful than PUBG.

By knowing the differences between the games, and learning how Fortnite actually works, you can be sure to collect the best gear and equipment. With a bit of reading you'll be kicking butt, taking names and you'll only get better as you learn many of the best kept secrets from some of the best players of Fortnite out there.

Be a player in charge of your own domain, learn what you need for success in Fortnite and become the skilled player that you've always wanted to be with this helpful getting started information.

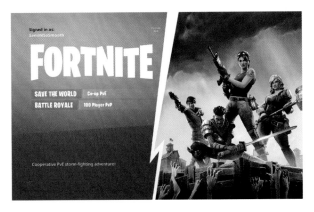

MISSION/OVERALL GAME STRUCTURE

Fortnite is structured much like a battle to the death and works in a way similar to Hunger Games if you're familiar with the books or movies. When you get into the flying bus, you have to be the one to choose a spot to land and fight your way to the end of the match. You cannot just hide to win the game though. You have to actually fight your way to the end. The map is considerably smaller than it is on PUBG, and you're much more likely to get into the action before the end of the round in Fortnite.

Your mission, should you choose it, is to be the last one standing if you're in that mode. There are three different modes. The only difference between them is whether or not you are alone. Some enjoy the solo modes, while others prefer people to pair up with.

Get on the bus, choose a destination, soar down to it and loot. Once you've looted, you run around, taking out people, staying within the storm's eye and making sure to stay alive. Each game depends on you strategizing your way to victory, but once you do, you will feel proud being able to make it out at the top over 100 other players.

Remember, you have to loot unless you're going to get everyone with the pickaxe that you start off with! Your first priority should be to pick up ranged weapons, shield and ammunition when the round gets started. These items are stored away in crates, hidden within buildings and in the backs of trucks. It's up to you to get moving quickly and to track down what you need to win each round. So get moving and track down the equipment that's going to lead you to victory!

DIFFERENCES BETWEEN SOLO, DUO, AND SQUAD MODES

There are a few different modes to be aware of when you're playing the game. Each one changes just a bit, but the game remains the same. You still have to battle your way to the end to win and you cannot just camp out.

Which mode you choose to play depends on whether or not you want to play alone or with others that are also playing to win. You can team up with those that you actually know or regulars you play around or you can have a random group of people playing chosen for you.

SOLO: Solo mode enables the player to play on their own without any teammates to work with to win. The player must make it to the top above 99 other players without any help.

DUO: Duo mode brings together a pair of players. They work together to bring the others down. They are then victors at the end together, not separately.

SQUAD : Squad mode is more of a team match than anything else. It splits the round up into two groups of 50 that have to overcome another group. Try out this mode if you want to learn what it means to work as a huge team, and experience the fun of overcoming another huge group of people. This is a good mode for making new friends as well!

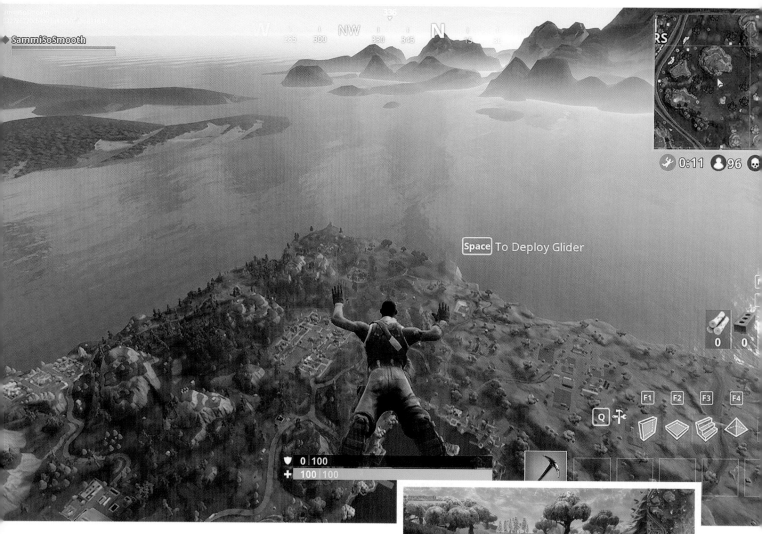

JUMPING FROM THE BATTLE BUS/CHOOSING WHERE TO LAND

Jumping from the battle bus is a must. You have to get off of it before it gets to the end of the map. It will warn you when it is about to do this. You can see all of the cities below since they are mapped out for you and each place has its name above it. You want to land in the area that you feel the most comfortable landing. Some of them have more loot, others have more shelter, many of them have more people. The choice is ultimately yours and it's a strategic one so start trying out spots and figure out which one is best for you!

Know how to get down there, know where to go and know which spot offers what. Having this information is going to help you out immensely each time you try to win at Fortnite. This expertise takes time to develop, but with experience you'll begin to get more comfortable with particular spaces and that's when you'll begin to enjoy success as well.

HOW TO GET TO GROUND FAST

Getting to the ground fast is something that a lot of people don't know how to do or realize is important at all. You want to use the glider as much as possible when you jump from the bus in the sky. You need to move yourself to where you want to go while also trying to drop quickly.

Depending on the console you're playing on, you can navigate your way by using the keys on the keyboard, the joystick on the console or even the mouse. You can then press down on these to move yourself closer to the ground and fall quicker.

To fall with as much speed as possible it's important to avoid popping your parachute until the very last minute. Instead angle your character down at the ground and drop with as much speed as you can. If you don't pop the parachute yourself the game will deploy it for you, so you never have to worry about crashing to your death.

Try to make it down there as fast as possible, especially if you're going to a well populated, high looting area. You will want to grab all of the good loot before everyone else so you're ready to begin the fight for your life.

PROS/CONS OF CROWDED AREAS VS. WILDERNESS

Crowded areas and wilderness spots with small shacks and cabins are both options throughout the map. When you are choosing the best place to land, you'll need to choose one or the other. Which one you choose depends on you and what you want out of each one. They both have pros and cons, which means that you have to either check out the pros and cons on your own or you have to try each one of them out on your own.

Crowded Areas: A lot of people choose crowded areas for one reason and one reason only, to get the best loot. When you get the best loot, you can ensure that you take out some of the other players, since you have a better gun, right?

If you're not one of the very first players to hit up these heavily looted locations, you're likely to get taken out. With any of the cities on the map there's a great opportunity to grab the best loot in the game, but the risk of being killed off early on is much higher than it is for players out in the wilderness.

One good strategy that many players use initially is to run around like wild while listening for crates as they go. They will sing to you and they are not necessarily in the highly crowded areas either. They are anywhere and everywhere throughout the map and they will change every time. You need to listen for them and then follow the singing to epic loot opportunities. They change the loot that is inside them, providing a lot of high quality choices that cannot be found in other areas.

Where you drop doesn't have any deciding factor on the class of items that are found there, since this is always a random thing that happens, including in the drops that are made. The only difference is that the cities have more loot, so the chances of finding better stuff is higher.

The Wilderness: Some like to drop into the wilderness where there are cabins and small homes to loot. These areas are more peaceful because fewer players drop here, though nowhere is safe throughout the entire game. You just have less looting, which means less people. You will have to run farther to avoid the storm in most cases when out in the wilderness, but this gives you time to look in all of the cabins and come up with a strategy while you let everyone else thin the herd during the initial chaos of the match.

Not everyone enjoys dropping into the wilderness though because you never know what you're going to find lootwise, and some players prefer to be in the thick of the action. That's why many of the players turn to more crowded locations around the map instead.

SOME OF THE BEST SPOTS TO LAND

Landing is a big decision to make. You want to land in the right area and depending on what your situation is and the skills that you have, you might want to take a different angle when landing. You should choose an area that you feel the most comfortable with and that is going to provide you with the most in terms of benefits and value for your survival in Fortnite.

Following are some of the best locations on the map to check out. Get familiar with each of these spots and you'll soon decide which you like and which are best avoided for your personal playstyle.

Keep in mind that the map has been updated and continues to be updated. While this is the most current map information now, it might not be in a few months since new locations are constantly being added, as well as changes in the places where loot and other items spawn.

▲ Haunted Hills & Junk Junction
(Top Left Hand Corner of the Map)

If you're looking for an awesome place to land that comes with some pretty awesome loot chests, then falling just to the right of these two places in the broken down houses next door is where it is at. You can find loot up to your ears hiding here.

Once you're done looting the houses, you can move over to the actual towns and find out what other loot (or players) are waiting for you there.

▲ Flush Factory & Those Shifty Shafts
(Between the Two Places)

If you want to gear up quick and have gear that actually gets the job done then this quiet area between two larger areas is where to go. A lot of people go to the larger areas, which makes it easier for you to survive in this less frequented location.

In here you'll find plenty of loot lying around, and there's an abandoned house with more guns for you to pick through as well. You will be set to take on the larger areas when you check out this smaller, broken down, quiet area first.

◀ Flush Factory 2
(To the Right of Original Flush Factory)

Overcrowded is probably a good word to use for this place — both players and loot. It is like everyone went crazy and wanted to give out a bunch of stuff to everyone.

This zone is great for wood from the pallets that are everywhere, but also for golden singing chests. In cars, in between buildings, on top of poles, you name it — listen and they will find you!

▲ Anarchy Acres (Motel Located Here)

The trucks and motel in this area all have golden chests to look out for, which makes this a must-see location on the map.

Go through the houses on the outside perimeter to find anything located here and then make your way to the hotel. You want more than your pickaxe when you make your way into the hotel itself since everyone likes to go to this area. Many chests are waiting with awesome loot, so it is important to go in locked and loaded for a fight for the good stuff!

▲ Tilted Towers (Motel Located Here)

Centrally located, if you want to be in the thick of it all but also have access to a huge amount of loot, then this is where you want to land. With crates everywhere and a pile of chests, everyone stops here on their journey through the land.

You want to keep in mind though, since everyone stops here, you will definitely have a fight on your hands so use the ample cover of the area to stay out of someone else's crosshairs while you gear up for the fight!

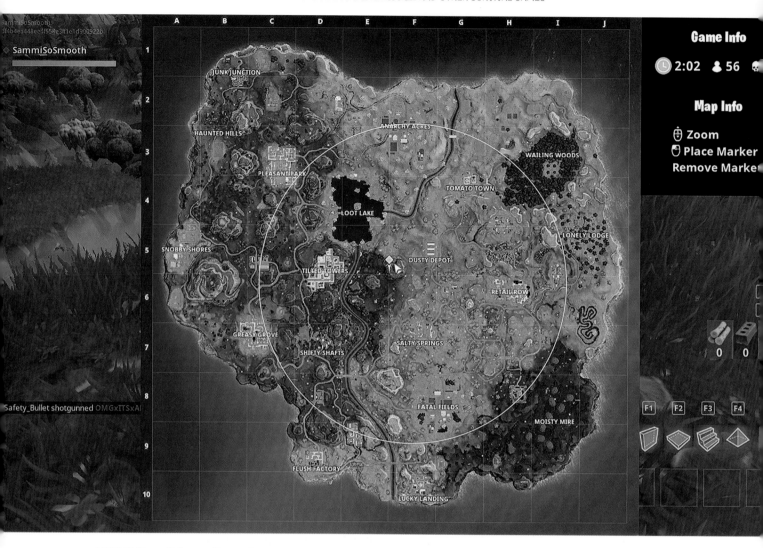

READING THE MAP AND STAYING WITHIN THE STORM EYE

As you play through round after round, one important task you need to keep up with at all times is staying within the eye of the storm. The storm is a very real danger, and it can take you out of the game before you even have a chance to encounter another player. The game gives a warning when the eye is about to move in, so you want to keep your eye out for this as you play. Once it does, check the map to see where you need to run and how far to be inside the eye. You'll have a timer for how long you have to make it to the new safe zone, move quickly but keep an eye out for other players waiting for you to run in.

Take a look on your map on your right hand side and you will notice that there is a white circle. If you stay within the white circle of the storm then you are safe from

any damage that might come your way outside of the storm. You should always make sure to refrain from being outside of the white, since the blue circle that you see on the screen is the storm.

The screen will also give a big warning every time the eye of the storm is about to move in on you. You want to make sure that you are paying attention to where the white circle is. If it is moving in on you, then you have a short time to get into a safe spot within the eye once again.

Always keep your eyes on this circle, as it continues to get smaller throughout the entire game. If you're left on the outside the circle then you will continue to take damage until you go to the inside of the circle. Eventually, staying outside of the circle will kill you.

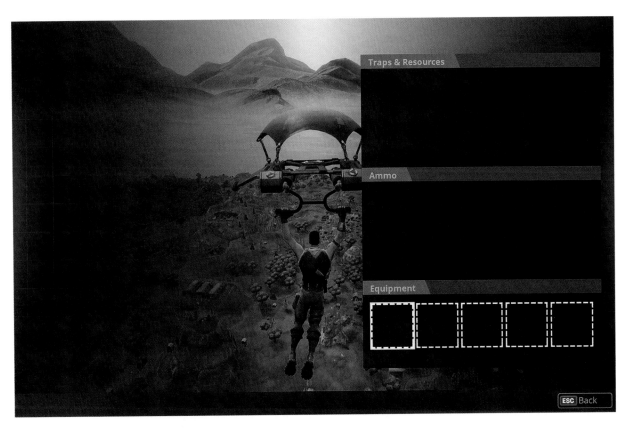

MANAGING INVENTORY/HOW TO ACQUIRE WEAPONS AND OTHER ITEMS

You have an inventory and you have to make sure that you acquire the right items as you go along. Many players go into Fortnite worried about holding too many items, with the belief that too much will slow them down like being overburdened does in other games. Unlike some of the other games out there, such as ARK, you do not have to worry about being weighted down, even if you're carrying around four trees worth of wood.

You have slots where your items but there is no quick way to rearrange items in slots, so it is important that you go somewhere safe while you go through your items and switch them around to make them easier for you to access while playing the game.

Bear in mind that you do not have a lot of space for everything. You cannot possibly keep it all with you, so you have to make way for more items by removing those that you do not need. Using some of the items, such as the shield potions, as soon as you get them can help you reduce the amount of spaces you use in your inventory for all of the items that are being found.

TIPS FOR MANAGING YOUR INVENTORY

Here are some pretty awesome tips for managing the inventory that you do have. Using these tips will help you become more efficient and lethal in each round of Fortnite, which will result in more wins for you!

Don't Hoard: This is a must in real life, as well. You don't want to have an overabundance of items that you have to get rid of. If you don't need it, don't take it or put it down. Keeping lower quality items or duplicates that you won't use offers you no benefit. Shed the excess and keep grabbing the loot that matters while ditching the rest.

ACQUIRING WEAPONS

Here are some of the best loot locations that are going to provide you with a way to benefit from finding those great weapons sooner, rather than later. Keep in mind that these places that have a lot of weapons and loot are also highly populated with players. You will have to fight them off to make sure to grab the weapons that you want and need.

Of course the tips up above are just to help you get started. It's up to you to choose your favorite spots, learn loot locations that you can go back to again and again and decide whether you like the popular locations or the less populated wild spawn points. You'll have to do some research as a new player, but over time you'll learn what you like and develop the skill to turn up loot when you need it most.

1 ABOVE THE WAILING WOODS — MULTIPLE FLOORED TOWER

A couple singing chests can be found in the middle and another right at the very top of the tower so make sure to check all of the floors to find them.

IN THE WAILING WOODS — ICE CREAM TRUCK

Check not only by the ice cream truck that spawns, but also inside the truck. Sometimes they leave a lot of goodies in this area to make use of.

2 PLEASANT PARK — THREE STORY HOUSE IN THE MIDDLE

If you want to find a lot of goodies the park has a bunch and so does the house in the middle of the park. It is definitely worth taking a look-see.

GREASY GROVE

You will find the most chests per square space in this little town. There are around 15 chests that spawn in the area.

LOOT LAKE

There are definitely chests filled with goodies around this lake, just make sure that you actually search out in the water, because that's where many of them hide.

SNOBBY SHORES

Perfect for finding 2 chests per house and not a lot of people drop in this area, giving you the freedom to roam throughout the area without getting caught or killed.

USING SOUND TO YOUR ADVANTAGE

Sound is a big thing when you play Fortnite, so you want to make sure that you have your speakers turned up and that you can hear what is around you. There are many instances when this would come in handy for someone that is playing the game and trying to get the most out of what is being offered.

Sound can tell you a lot about your surroundings — for example, you will be able to hear when there is a firefight happening right next door or when you are being followed by someone. You will hear the rustling of the grass as they step through it or their footsteps on the wood floor.

You can also hear the crates singing to you as you get closer to them. The closer you get to them, the louder the singing ends up being.

Those that are playing without sound won't have the ability to hear when someone is close to them, even if you are the one hunting them down. You want to be able to know what is happening around you and just looking is not going to help you. You need to be able to hear them too.

Playing with sounds will help you be a better player overall and ensure that you can find what you're looking for when playing the game.

SOLO GAME MODE TIPS AND TRICKS

· Choose the best landing spot since this is going to set you up for success early on in the game. You want to go somewhere that you feel comfortable, whether it means getting more loot at a highly populated area or laying low in one of the far off areas.

· Grab all of the ammo from the ammo boxes that you come across. These are vital to your success throughout the day and you need to stock up.

· Always grab materials and trap schematics that help you build and craft everywhere you go. Do not underestimate the ability to use these items while you are out and about. If you don't have enough materials on you to build something then this can be a problem.

· One of the best things that you can do to win the game is to perch yourself at high ground. You want to be above everyone at a place that you can see everything below you. Getting high will help you stay towards the lead.

· When you're in the middle of a duel, make sure that you keep a level head about what is happening. Make sure that you know the best possible place to put yourself when it comes to fighting someone. In order to win you have to have the best advantage, which means you have to think the fight through.

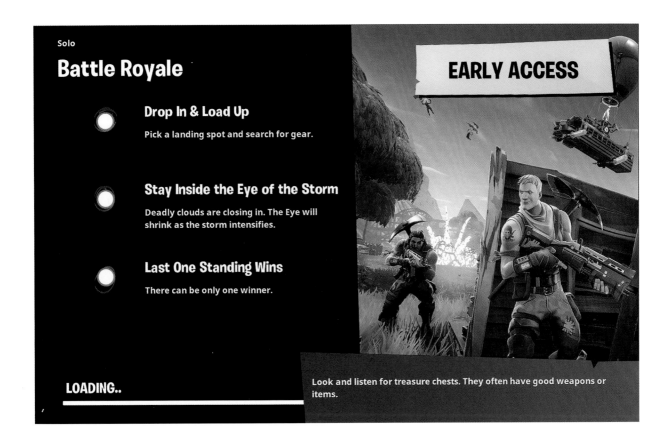

Solo

Battle Royale

Drop In & Load Up
Pick a landing spot and search for gear.

Stay Inside the Eye of the Storm
Deadly clouds are closing in. The Eye will shrink as the storm intensifies.

Last One Standing Wins
There can be only one winner.

EARLY ACCESS

LOADING..

Look and listen for treasure chests. They often have good weapons or items.

· Always keep your ears open while looting or doing anything else. You want to make sure to listen for approaching footsteps. They might want to take you out when it comes to grabbing items that they want or just to get you out of the game and put themselves at a further lead.

· Be wary of open doors in every building that you go through. They all start closed and if they are open then someone opened them and left them. This could have been because they were in a hurry and are just passing through or they could just have set out a trap for you to go through. You never know.

· Always play around the storm. You want to either go around the outside of the perimeter of the storm and then move in when it moves in and follow it or play just inside the circle of the storm as this is one of the best

ways to make sure you're always inside it but not in a common area that you will be taken out quickly in.

· Shield potions are a must for anyone, in any of the game modes. You want a shield that is going to protect you much more than a regular one would. You want to take it as soon as you get it since it is going to offer you the most protection, even if you are not currently in the middle of a fight.

· Always check the weapons rarity and get rid of those that are lower if you find something better to work with. This is going to give you a much better advantage and free up some of the space in your inventory that you could use for something even better.

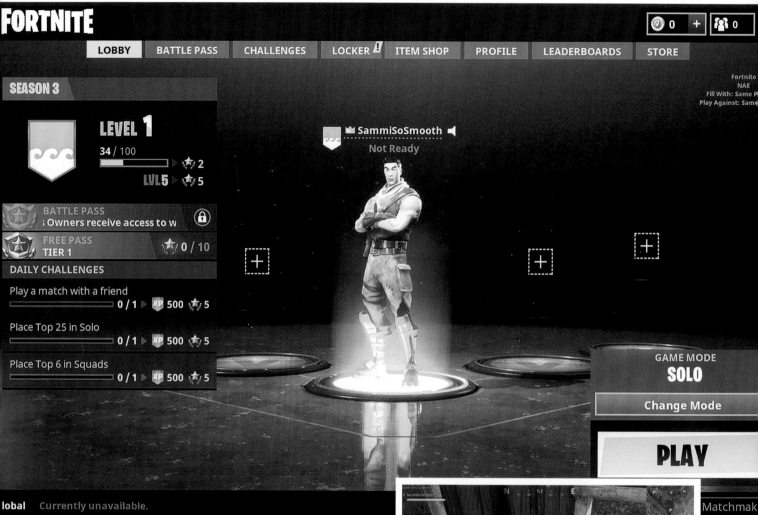

BIG PICTURE STRATEGY

Everyone is going to have something different to say when it comes to the strategy that you use to win the solo version, as well as the other versions of the game. That is fine and you might want to take some of what they say to heart, as many of them are great tactics to help you make it to the top. Fortnite can be difficult to navigate and actually win in the beginning so becoming familiar with the game is the best thing to do.

First things, first.

You will want to go into your settings and set them in a way that you feel comfortable playing. This could be changing the sensitivity levels to the screen adjustments

to the controls that you use. Getting your game set up so that it's familiar will help you as you begin playing through rounds of the game.

You will then want to practice shooting. This is going to be something that is hard to do at first but again, since the guns just need to be pointed in the general direction, unlike PUBG, you will have an easier time shooting at the

target and actually hitting it. You can practice right in the battlegrounds before going out to the actual game, but one of the best ways to get practice in is to try highly populated areas, grab some guns and get fighting!

Make sure that you know where to land, how to land and what to do next...

If you're new to this, avoiding those popular, densely populated landing sites is the best thing for you right now. You need to get your own game plan together and you're not going to be able to do this when you land right in the thick of it all. You have to refrain from doing this. You need to work with what you have, find some supplies and then come up with a way to make it to the top.

When you do land, make sure to drop down as fast as you possibly can. You want to be the first one to land in that area, take what you want and then get out of the area before someone else comes and takes the items. You need to make sure that you do this quickly every single time you play.

Once this is done, you should move on to the next area and find cover once you're equipped to take someone on. You will feel much safer when you have a weapon on you and can defend yourself. It is definitely a game about survival so you will want to make sure that you are protecting yourself at all costs. It is either shoot or be shot - it is as simple as that - so you will want to become pretty good at shooting in the game. That's why practice is ideal.

ALWAYS avoid the storm. You will end up hurt or dead if you are outside of the white circle when the storm comes and causes problems. You do not want to end up out there when it is moving in. You have a specific amount of time to get back into the white circle but once that time is up, if you do not make it, you are going to be a goner and it will kill you and kick you out of the game. Make sure this does not happen by being in the circle when you need to be.

Listen to your surroundings closely. Even though we mentioned this above, this is actually one of the biggest tips out there that you should make sure to take advantage of. You want to make sure that you are listening to each and every creak, crack, groan or swish that you hear. This will make a huge difference for you with each round that you play through, so get used to listening now.

Never, ever shoot your gun for fun. Guns in this game are big and cool but if you're not into having someone find you, especially if you're hiding, then you want to keep your gun shooting to a minimum. If someone hears you, they will come running and then it is going to be a duel to find out who is going to make it to the end. You never know who might come out last, it might not be you, so it is wise to keep your gun ready without using it until you really need it.

There are many ways to win this game and many tips and tricks that can help you, but you should always make sure to keep the big picture in mind. That is to make sure that you kill or be killed. The main objective is to survive to the very end and that means taking out any players that you may come across. If you don't take them out, you'll have to worry about them coming for you, and 9 times out of 10 that's exactly what's going to happen. Be smart, take out targets that you can whenever possible and you'll grow in skill while having a better chance of becoming the number one player in a match.

When you're ready to play, prepare yourself and try a few rounds. Practice makes perfect and you will get your own strategy down to a T. Make sure to use the game tips and strategies that are out there to make the most

of your game time. You don't want to worry about not being able to make it to the end. It will take you some time but it can be done.

Have fun on Fortnite and remember, you will become a victor when you have the skill and experience to take on everyone else, so get practicing! ⭐

WEAPONS

You can increase your chances of survival when you have a good idea of where each of the weapons is hidden, their powers and what they are. With new weapons showing up every so often, there are many of them to be found but also many more that might be added in the near future.

The weapons are what are going to help keep you alive as you make it towards the end. You cannot just hide out and wait for something to happen, you have to fight to win and with these weapons, you can easily do so.

Not only that, but using some tactics to go along with these weapons can really make a difference when trying to win. The weapons are just a part of the game, your strategy is going to make a huge difference on whether you succeed or perish with the rest.

COLOR CODES

The colors on the weapons mark their rarities and which weapons are more commonly found than others. There are five different rarities. You might even find the same gun but with different rarities, the better the rarity color of the weapon, the better the stats on the weapon.

From strong to weak, the colors are as follows:

● ORANGE: LEGENDARY
● PURPLE: EPIC
● BLUE: RARE
● GREEN: UNCOMMON
● GRAY: COMMON

TYPES OF GUNS

There are many types of guns that are in the game and going from the most rare to the most common, here are the guns you can expect to find and some of their stats.

ASSAULT RIFLES

GUN	RARITY	DPS	DAMAGE	FIRE RATE	MAG
Scoped Rifle	Purple	84	24	3.5	20
Scoped Rifle	Blue	80.5	23	3.5	20
Burst	Blue	121.9	30	4.06	30
Burst	Green	117.9	29	4.06	30
Burst	Gray	109.7	27	4.06	30
SCAR	Orange	234.5	36	5.5	30
SCAR	Purple	218.5	35	5.5	30
M16	Blue	192.5	33	5.5	30
M16	Green	181.5	31	5.5	30
M16	Gray	165	30	5.5	30

PISTOLS 🔫

GUN	RARITY	DPS	DAMAGE	FIRE RATE	MAG
Pistol	Blue	168.7	25	6.75	16
Pistol	Green	162	24	6.75	16
Pistol	Gray	155.2	23	6.75	16
Revolver	Blue	54	60	0.9	6
Revolver	Green	51.3	57	0.9	6
Revolver	Gray	48.6	54	0.9	6
Hand Cannon	Orange	62.4	78	0.8	7
Hand Cannon	Purple	60	75	0.8	7
Suppressed Pistol	Orange	189	28	6.75	16
Suppressed Pistol	Purple	175.5	26	6.75	16

SHOTGUNS

GUN	RARITY	DPS	DAMAGE	FIRE RATE	MAG
Pump Shotgun	Blue	66.5	95	0.7	5
Pump Shotgun	Green	63	90	0.7	5
Tactical Shotgun	Blue	111	74	1.5	8
Tactical Shotgun	Green	105	70	1.5	8
Tactical Shotgun	Gray	100.5	67	1.5	8

SUBMACHINE GUNS

GUN	RARITY	DPS	DAMAGE	FIRE RATE	MAG
Minigun	Orange	204	17	12	0
Minigun	Purple	192	16	12	0
Tactical	Purple	234	18	13	35
Tactical	Blue	221	17	13	35
Tactical	Gray	208	16	13	35
Suppressor	Blue	171	19	9	30
Suppressor	Green	162	18	9	30
Suppressor	Gray	153	17	9	30
SMG	Blue	240	16	15	35
SMG	Green	225	15	15	35
SMG	Gray	210	14	15	35

SNIPER RIFLES

GUN	RARITY	DPS	DAMAGE	FIRE RATE	MAG
Semi-Auto	Orange	79.2	66	1.2	10
Semi-Auto	Purple	75.6	63	1.2	10
Bolt Action (AWP)	Orange	38.3	116	0.33	1
Bolt Action (AWP)	Purple	36.6	110	0.33	1
Bolt Action (AWP)	Blue	36	105	0.33	1
Hunting Rifle	Blue	72	90	0.8	1
Hunting Rifle	Green	68.8	86	0.8	1

OTHER WEAPONS

If you're looking to find a gun that is unlike your normal ones on the above list, then you have come to the right place. With rocket and grenade launchers, as well as a couple of other assorted weapons, you can be sure that you have the stats and know which other guns are out there to make use of.

ROCKET LAUNCHER

GUN	RARITY	DPS	DAMAGE	FIRE RATE	MAG
Rocket	Orange	90.7	121	0.75	1
Rocket	Purple	87	116	0.75	1
Rocket	Blue	82.5	110	0.75	1

GRENADE LAUNCHERS

GUN	RARITY	DPS	DAMAGE	FIRE RATE	MAG
Grenade	Orange	110	100	1	6
Grenade	Purple	105	105	1	6
Grenade	Blue	100	100	1	6

CROSSBOWS

GUN	RARITY	DAMAGE	SHOT (INFINITE)
Crossbow	Purple	79	1
Crossbow	Blue	75	1

GRENADES

GUN	DPS	DAMAGE
Grenade	210	105
Boogie Bomb	0	50%
Impulse Grenade	0	50%

COMBAT TACTICS

Combat tactics are important to make note of while playing this game, because you want to stay alive. The only way this is going to happen is if you have the best tactics to help you do so. With these tactics, you can try to stay alive for a longer period of time, while also working on ways to overcome the mission and take on the champion level of the game.

1 Choose Landing Points Wisely

Jumping in an area that has a lot of loot, is wide open and where many other people are jumping into is a sure way to lose your life 99 out of 100 times. This is why it is important to choose your landing points wisely while playing.

Try to drop towards the middle or far edge of the map, or even as soon as you get on. When you do this, you have a better chance at finding loot and not being in the thick of it all where everyone else is going to drop and just get to killing straight away. Think wiser.

2 Use Building Structures

When defending yourself, make a wall that cannot be penetrated easily. Quickly load up the crafting menu and throw up the wall or create a shelter, a maze or other building structure that would be hard for the enemy to get through and gives you enough time to get a plan, load your weapon or just get away for a short period of time to heal and survive.

These building structures are a great way to get some down time and save your life. It has been a proven method for many when they are feeling like they are slipping behind or may die.

Listening for Sounds

Almost everything in Fortnight gives off a sound that lets you know that it is near or that you're coming near something. It is important to keep your volume up and to listen closely.

Sprinting, from far away, is one of the best telltale sounds to listen for. This means that someone has scoped you out and they're coming to attack. You want to make sure to get out of there and FAST! You could also stay and fight if you have the weapons you need to do so.

The golden chests that you will find loaded with some of the best loot sing to you when you get near them. If you listen closely, you might be able to find them easier and get to them to loot them. If they're singing to you, now is the time to move and discover them before someone else does.

❸ Don't Loot Bodies Straight Away

It is very possible that someone is sniping that area and you do not want to be the first one to get picked off when it happens. You are at the most vulnerable point in the game while looting bodies or homes. You're paying attention to what you're doing and not to your surroundings, making it easy for those around you to pick you off if they need to.

Building a wall around you while you loot and try out gear is also a great way to protect yourself while you are vulnerable. If you are not paying attention to those around you, always have a wall built.

Launch Pads Are a MUST

In this game, playing means winning and when you're trying to win, you need something that is going to save your life in a pinch. A launch pad can do just that. You can throw the pad down on a flat surface and launch yourself up into the air and move to a completely new area. It is almost like being able to fly away during a fight that is leaving you as the loser.

You can even run away from the blue zone when the circle starts to move in too quickly or when you need to run away from a group of people that are ganging up on you. There are many valuable uses for launch pads and they might even help you win the game.

4 Hold the Pickaxe to Loot Faster

During combat, you have to act fast, and the same goes for looting. You want to make sure that you pick up everything without having to rearrange it all when you do so. If you're holding your pickaxe then this does not become a problem. Everything will easily slip right into the necessary place that it needs to be in and you can rest assured knowing you have all the ammo needed to kick some butt.

There Really Isn't a Bullet Drop

In some of the other games out there where you have to fire the gun, you have to fire above the person's head to have it move down slightly to hit the person where you need it to hit them. This is known as bullet drop. This, however, does not happen in Fortnight except for with the sniper rifles.

Every weapon is hitscan which means that you can shoot at the enemies and you don't have to shoot directly at them or over them to get a hit. It makes fighting easier but it also means that you can get taken out by someone that is not so great of a shot, as well.

5 Be Flexible

When you're trying to win the game, then you want a tactic that is going to make sure that you have a way to go through the different steps and make sure that you win in the end. Using one of these tactics, or even trying them all out for yourself can give you a way to benefit.

Each one of these tactics does come with pros and cons, so it is important to find out which one you feel the most comfortable using while playing.

6 Staying on the Edge of the Storm

Staying on the edge of the storm consistently is one of the strategies that top players have used in the past that feel that it works for them. This then ensures that you move only when you have to and it helps you keep an eye on your surroundings at all times. It puts you in control, is slow moving but also reduces the chances of getting shot in the back.

7 Build a Sniper's Nest High Up

If you're all about taking people out and you don't really want to move anywhere then this is a good strategy. You will need to build your own nest that is hidden from the rest of the people, find your rifle ahead of everyone else and have a decent amount of ammo. Sit up there and pick

off people one by one that are advancing in the game. This gives you a higher up position that you benefit from and it might just win you the game.

8 Scouting Heavily Scavenged Areas

If you're looking for a way to take out multiple people in a short period of time then this is the best way to go about doing so. You can sit and wait at these areas where people are found going through crates and other areas and take them out. You have to hide and watch for them but once they come, you can use any sort of fighting tactic that you like.

9 Waiting by the Safe Zones

When the storm starts to shrink, you will see where you need to stand. If you're sitting in the right place, you will watch people run into the safe zone to get away from the moving eye. This gives you a perfect opportunity to take a lot of people out while you sit there. Again, you will want to stay out of sight while you do this, since it is going to be easy for them to see you while they approach if you don't.

Be Confident

Don't let them know your weaknesses and just play the game. Newcomers find it intimidating at first because you want to succeed but you also know that it is hard to do when you have no experience in the game. If this is the case, scope out the area a few times and know it is okay if you don't make it the first few times.

Once you become more comfortable and acquainted with the game, you will then be able to take advantage of the spots in the game, specific locations and even ways that you can build up a game plan or tactic that truly works. Grabbing the trophy for winning these matches is the ultimate thing, especially since you start with 100 people and you are the last one standing — it is a pretty big achievement!

10 Don't Engage Unless You Are Sure
You want to survive, that is the goal of the game. This means you don't want to engage until you are certain that you are going to win and they can be picked off easily. If you choose the wrong battle, you might be the one that loses. This is definitely not the way to win the

game. Some come into the battle with a certainty of winning, only to find out that they cannot go around getting kills and win the game.

The game isn't about the number of kills you get, but the amount of time you're able to survive while playing it. If you can avoid confrontation, then you're able to survive the next few rounds that come about.

11 Find a Shield Potion? Drink It Immediately
If you find a shield potion, you should make sure to drink it immediately. This is a must because you never want to worry about not being protected. The best part about these potions is that they are able to protect you for the duration of the game. This is a big perk when trying to make it as far

as you can. While they cannot protect against all damage dealt, they can definitely protect against most of it.

Many people feel that hiding until the last possible moment is the best way to win the game, but this is not true. You can hide out and be one of the last 20-30 left in the game but you have to be able to take on these people when you go to fight them. You have to be able to win. If you hide the last bit of the game, you will have a hard time having any materials, weapons or anything else to have a plan on winning with. You may have just wasted a lot of the game hiding in the structure you built or found.

Remember while you're playing that there is a difference between staying alive and winning. You cannot just hide out and win the game. There is always going to be one

winner and you have to find and remove the other person to be that winner.

Starting in the right areas is going to help you keep afloat when those around you are getting picked off. If you're new to the game then you obviously don't want to start where there are going to be many people jumping off. They will pinpoint your location and once they have looted the area, they will take out those that they come across on the board.

Come up with your own tactics to fight off those that you're playing against. You want to make sure that you do what is comfortable and if you're new to the game then it is important to try a few different tactics before going with just one.

OTHER LOOT

Preparing yourself ahead of time to grab not only the most loot, but the best loot is always a good thing. When you drop from that bus, you know you have seconds before someone sneaks up on you that has already found loot. They're waiting to take you out.

This won't happen if you get to the crates, grab your weapons and make sure to know where the best loot is hiding. It is also best to learn about some of the best places to land in the game, since this can also put you in the best spots to grab some loot and run.

Even though most of the best loot is found at the upper portion of the island, there are still a few spots below that you will want to make use of when trying to grab the loot and run.

Never let them catch you!

CHESTS

Chests spawn in various locations throughout the game. To grab some of that sweet chest loot you need to make sure you know where chests spawn and that you get there before other players do.

It is going to be hard to fight off others from the chests if you have nothing to fight them with. Get there quick, be fast and don't be seen!

Chests randomly spawn throughout the map, so it is important to know that even though some of the areas might have reported having items in them at this locale, it does not mean they will spawn in that area with those things again.

Greasy Grove has the most loot of all but it's a really dangerous part of the map. Not only that, but if you drop in some of the older areas, you will find that there is a decent amount of loot and less players that are dropping in that location. They're going to the newer cities on the map.

The chests can hold anything from shields and weapons to health potions and healing supplies. You can find traps in them and other special items that are going to help you stop people from getting you while you're playing the game.

SECRET CHESTS!

There are actually many secret chests that a lot of players do not know about. They're hidden in areas where people do not look. One of the biggest places to look for these secret chests is on the side of cliffs. They are hanging there and allow you to open them and check out the loot inside.

Over the hills and in hidden alcoves inside the sides of them are hidden chests with loot. These hidden areas have many great items inside them. Hidden chests inside barns are good to look for.

Remember!

Keep your ears open to listen for these secret chests! They will sing to you when you come near them so you want to make sure that you can hear them when you're running through the area.

RECOVERING HEALTH

There are many ways that you can recover health while you're in the middle of the game. You will want to do this after you have taken cover in a building, built your own shelter or found an area under a hill to hang out under while you heal yourself.

Healing items can be found inside chests, on shelves and in different areas throughout the game. They are lit up and when you mouse over them, you will find that you can easily pick them up and add them to a quick hot button to easily use them as needed.

◀ SHIELD POTIONS

There are two types of shield potions (large and mini) that you can use to protect yourself during game play. While these don't actually heal you while playing, they will create a shield, absorbing any damage that comes from the outside.

The mini adds a 25 buffer to your health bar, while the large adds 50. They are arguably one of the biggest and best loot pickups of the game. The mini ones are smaller and a bit easier to find than the larger ones, so you can carry a couple with you and top off as needed.

◀BANDAGES

Bandages are great to have and you can find them in stacks of five, but they will not restore your health all the way. You can only heal up to 75% with them which is better than nothing when your health is super low. Each bandage restores 15 points of health and take a minute to bandage up and heal up the whole 15 points.

◀SLURPJUICE

This is juice that you can chug right before a fight to not only protect yourself but also boost your health as you're fighting. You get 25 armor, like from the mini shield potion, as well as 25 health and this is given over a period of 25 seconds. Fight away and stay protected and healed the entire time.

◀MEDKIT

You can only carry around three of these on you so make sure to keep them on you, use them wisely and always have at least one. They are one of the two things that is going to boost your health up to 100 max points.

◀CHUG-JUG

These are the second things that are going to boost up your health to the max. They take a good 15 seconds to fully down but once you do, your health will go back up to where it needs to be. You also get the most out of the shields you have on, as well. Double the greatness in a drink!

It is always good to manage your resources and that means all of your healing items. You never know when you might need them while you're in the middle of a fight. They can prove to be the one thing that is needed when you're in a pinch and need to make your way out of it.

TRAPS

Traps are another beneficial item to find and use. However, learning how to use them can prove to be tricky, especially since you have to make sure that they are not seen when you set them up and learn how to do so quickly.

◀ DAMAGE TRAP

The damage trap has three different types of traps that it can turn into: the spike trap, wall zapper or ceiling zapper. Placing it in one of these areas will turn it into that specific item.

You just have to set it up and leave it. From there, it gets its victim if they are unsuspecting.

◀ LAUNCH PAD

If you're looking to get away from an opponent or if you want to launch someone or something into the air, then this is the pad for you.

Set it up and then bounce away. You can also set it up so that other players accidentally jump on it and are launched, as well.

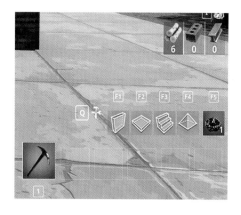

◀ COZY CAMPFIRE

This is usually looted as a rare quality item, so it can be quite hard to find. Once it is placed on the floor out in the open it is able to heal anyone nearby 2 points per second. It is pretty useful if you have a group or if you need some extra health and are alone.

All of these items have their pros and cons and sometimes they can be hard to find. You might have to keep your ears open to listen to a singing chest that is somewhere throughout the map because you will have a good chance of finding a trap inside one of those!

BUSHES

One of the neatest things about this game that a lot of players do not know about is the fact that you can seamlessly become one with a bush if you obtain the Bush Disguise. You will blend in with it and become a piece of shrubbery yourself. This is something that will provide you with a little extra time and also the ability to scout out some players as they run past you.

Keep in mind, you will not be able to keep blending in as a bush if you take any damage from players that might be in the area. If they shoot your way and they hit you, you will no longer be camouflaged into the shrub.

They are a consumable and you just eat it and crouch down and it takes effect within seconds. It is definitely a good way to save your butt if someone has spotted you!

WOOD, METAL & STEEL

We will go into more detail about these items that you can collect and carry on you further into the book but it is good to have some of these materials available as you run around. You want to be able to craft items and this is one of the only ways you can do so. In order to craft them, you need to be able to have the materials to build them.

Stone comes from the rock walls, the large boulders and other rocky areas. Wood can come from crates, wooden buildings and trees. Metal can be gathered from metal bins and canisters that you find inside the shelters.

This can become repetitive but it is worth it if you're going to be building a lot throughout the game. You need to be able to boost your resources and this is the best way to do so.

THEMED LOOT

During the many holidays that happen throughout the year, there are themed loot boxes and crates that are hiding throughout the game. This is done to add extra fun and a little bit of value to those regular players that come in to play.

Easter eggs are one of the favorites and during the holidays, you can expect to find nicely wrapped presents or even lights that show up on the bush when you become one. These are all nice additions that you would not expect to find in a deathmatch style game but they change it throughout the year to go with the specific holidays that are coming up.

We have yet to cross some of the other holidays out there, so we have to wait and see what the game makers have in store for those ones and what they might offer for an added bonus.

NEW RELEASES

New releases of items happen all the time. If you're looking to personalize a character and make them your own and you pay for Fortnite Battle Pass, then you're able to unlock a whole store full of items that you can use to your disposal. More is on this topic later in the book, but it is always good to read the beginning opening of the game to find out what has changed with it and see if they have released any new weapons or loot in the game.

THE GOLDEN UMBRELLA

Everyone wants the golden umbrella. This is what is going to help you glide down to the next game play in style. However, you cannot just go out and get the golden umbrella. You have to actually earn it or become lucky - or a bit of both during the game.

When you are able to come out as the victor in the match, you get the golden umbrella as a trophy for doing so. If you win a golden umbrella in SOLO mode, you will have to play the other modes to win the umbrellas in those since they do not transfer from one mode to the next. This is something that many people do not know about but it is important to keep in mind if you're trying to get a golden umbrella to show off in all of the game modes. It won't happen until you were a victor in all of them.

Loot is one of the most important aspects of Fortnite. Understanding how it works, where to find it and how to make the most of it will help ensure that you have a decent chance at winning every time you play.

Just keep in mind that you can only carry a limited amount of loot at once. You want to carry the most important items though and stay away from those you do not need. You only have so much space to carry the items with you. Most players stock up on the items that they value most and forget about the rest. Health consumables and shield potions are both very important to have while playing the game, so always prioritize these along with some key weapons for the best results.

BUILDING

Building is an essential task in Fortnite. It will save your life when you're in a pinch, you can create traps that lure other players in for easy eliminations, or you can create the cover you need to survive when there is a large firefight going on.

Players benefit from the structures they can craft so knowing what you can make, how to make them and how they might benefit you will ensure that you play more effectively and start beating less experienced players in the game. Before you know it you'll be the last man standing in your survival game and it's all going to be because you understand building mechanics.

Knowing more about the different elements and materials in the game can help you win more matches overall, especially if you are fast at putting up complex structures. Getting fast building mechanics down makes it possible to get a heal off when you need it, hide from other players, trick them into thinking that you're not there and more. You'll be amazed at all the things you can accomplish when you understand how to make the most of building in Fortnite.

DIFFERENT STRUCTURAL ELEMENTS

There are different structural elements that you need to know about in the building system. It's important to use the right elements at the right times as you play to be the most effective player that you can be.

All of the below structural elements are built using basic wood that can be gathered from different areas, so once you have gathered the materials, you can construct.

- **Walls**
- **Floors**
- **Stairs/Ramps**
- **Roofs**

DIFFERENT BUILDS AND CONSTRUCTING THEM

During the game, you want to make sure that your building is on point. You need to make different structures to help you accomplish the things that need to be accomplished. That's why it's helpful to understand the building system fully and how to use it when you need it.

Below is a list of basic builds that can be done by anyone. They're each good for a different situation in the game, so become familiar with each structure type and you'll be ready to handle more situations.

Here are some great pointers and different builds that will help you out of those sticky situations...

⬆ Panic Walls

You can place these walls in front of you when you're trying to create a shield between you and an opponent or just in the general direction from where bullets are coming from. This is always good to do when you're running. You can add walls as you go to slow down the enemy and give yourself some cover. Choose wood as the material and then place down walls in front of you. Wood will not break down as you are building it on the go.

Panic Ramps

If you're panicking and you need a way out, then these are the ramps for you! You can not only hide under the ramp but you can enjoy the advantage that higher ground provides you with as well, easily picking off opponents from up high before they even see you! Add a ramp and continue to add ramp pieces until you feel that you are high enough to where you want to be. Keep in mind, the ramps that are built are easy to break down and knock

through so you will want to make them quick and make sure your enemies are not right on top of you because they will break them apart.

You can further enhance your panic ramps with protective side walls as you get faster and more proficient at building. Just make sure that you're using the right materials for your walls or your defenses aren't going to do much for you. Wood is easy and quick to build with but offers little protection. With steel walls, you have a better chance of not being blown away and you'll give yourself more time before the enemy can reach you, letting you get off a heal or prepare a shotgun to fight back.

⬆ Sniper Tower

Building a sniper tower is going to give you the ability to see over everyone else and focus down on them, making it easier to pick off targets effectively. Creating the perfect tower is an art form because if you put it in the wrong location you'll be more likely to get knocked down off it to your death. That's why you need to choose your location with care for the best results.

When you build this tower it is extremely easy but make sure to put a ramp on one of the sides since this is going to be your quick escape, should you need one while you're up there. You first want to build walls around yourself and then a ramp in the middle. Go to the highest point of the ramp and then create more walls. Do this time and time

again until you are high enough up off the ground and can see those that are far away from you. You want to have the best sight so keep going until you feel like you're seeing opponents before they can see you.

You can even keep building onto your sniper ramp to create a sky fortress or to go into the tallest points of other buildings that might be in the area. The sky is the limit when it comes to a sniper tower that you build and make use of. Just keep building onto your tower to make it more functional as necessary while you play. Creating a sniper tower is one of the best things that you can do when there are minimal people left and you want to take them out first before they see you. You'll have the sky advantage for an easy victory.

WHEN AND HOW TO USE THEM

It's important that you know how to make the most of these structures when you're in the middle of game play. If you know how to utilize them at the right times then you can benefit yourself and win more regularly. With the skills of building quickly you are able to do many things that might otherwise mean you wouldn't make it as far as you do with the building being done.

Here are some of the reasons why you would want to build while in the game...

- Build up a wall when someone is shooting at you to provide further protection

- Use stairs to cover ground more quickly

- Use stairs to look ahead of you by making yourself taller

- Add a structure around you so you can heal up without someone sneaking up on you

- Ramps provide not only more height to see, but also cover when it is needed and a way to get into a building that you cannot climb up into

- Create a trap that lures potential opponents to their death so you can rise above

- Quickly build a building around an opponent so they're unable to move

There are many reasons to build in Fortnite and depending on the outcome that you want, you may find that one way to build is better than another. The best thing to do is to practice building so that you're able to do so quickly in a time of need.

BUILDING MATERIALS (WOOD, STONE, STEEL)

There are different building materials that can be harvested and used when you are building. In order to use them, you have to make sure that you have enough to create the specific plans that you want. The crafting will not work if you do not have enough in your inventory to build with. While playing the game you'll notice the list of materials to the right of your screen. You can check out all of the material that you have on hand at any given time with a quick glance to the side.

There are three main materials that you will be able to find and harvest for building in Fortnite. It's a good idea to have a bit of each of them on hand, though wood is the one that is used the most in the game. When you have wood, you pretty much have endless possibilities for structures that you can build, and you'll have the freedom you need to make good decisions as you play.

While gathering materials it will be up to you to decide if you want to focus on wood, stone or steel. They're available all over the place in Fortnite, so think about the materials you need and stockpile what you can as you go through the game.

HOW AND WHERE TO HARVEST

It is good to know where to go to harvest these materials and how to do so. It is important to note that they can be found almost anywhere, so you never have to worry about running out of supplies as you move from one area to another.

Make sure that you're ready to harvest using your pickaxe in the beginning of the game. You do not have to worry about having guns or anything else in the beginning because you start off with a pickaxe that allows you to gather materials as needed.

It can also come in handy when you do not have bullets and need to whoop some butt, too... so always keep that pickaxe handy!

Here is where you can harvest the materials from within the game. You just have to hit the material with your pickaxe to break it down and add it to your inventory. Try and remember to gather materials as often as possible while you move through the world so you're ready to build structures that you need when in combat.

Keep in mind that when you want to grab materials, you can also get them from doing expeditions and through reward packs that offer a bit extra when they drop.

WOOD	STONE	STEEL
Trees	Walkways	Metal Bins
Structures	Large Boulders	Containers
Boxes and Crates	Sides of Cliffs	Shelving Units

PROS AND CONS OF DIFFERENT MATERIALS

Different building materials have different strengths and weaknesses. Make sure you understand these differences so you can make the right material choice each time you build.

Many players choose steel or stone to create a home base out of when it comes to having a place for everyone to come and hang out until the end of the game. If you're playing by yourself, then it is important to have somewhere that you can run to heal and to ensure that you are protected from other players. Of course, your home base also has to be somewhere you remember and it has to be hidden from other players. This can be a bit tricky to accomplish.

Every build that you do will take up 10 of the material that you use. You will need to have 10 for a wall, 10 for a roof, 10 for a slope and so on.

WOOD	STONE	STEEL
Wood is the quickest material that you can assemble and put together so if you're looking to speed build, then wood is the way to go.	Strong material that is semi-quick to throw up and use.	Steel is the strongest material that you're going to find, which is a given so you will want to think about this when it comes to making a fortress that is actually going to protect you for some time and not get knocked down in a short period.
Wood is great to trick other players since it creates a weak point in a structure that could take them out.	It is aesthetically pleasing to the eye, which is something that some like to look for when putting up a building.	Steel is also very easy to find so in addition to it being strong, you will not have a shortage of it.
Wood is also great for putting it where people spawn in, so that you can give them a weak point to fall through.	Stone is harvested at a medium rate, so it should be used less often than wood is.	Steel is harvested at a much slower rate than wood or stone, so take care of this precious commodity.

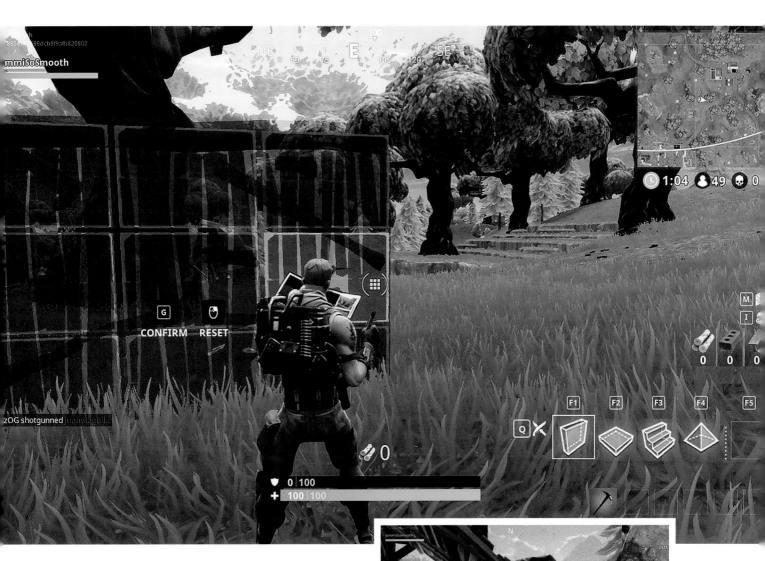

HOW TO BUILD

If you're wondering how you can go about building these structures, then it is important to use these methods for building. The method you choose depends on the type of console that you are playing on or if you are on the PC.

Building on the PC

You use the build menu that is accessible at all times throughout the game and on all of the consoles that you can play on. This can be done by pressing Q or any of the F1 to F5 keys on a keyboard. G edits the materials to a different angle and R rotates it so you can put the material any way that you would like. These controls can become almost second nature.

Building on the PS4

PS4 toggles are a bit different. You need to press the circle to open up the building menu and then hold it down to edit your build. The triangle cycles through the menu for you, while the R1 can rotate the structure that you are building. R2 places the structure that you choose and L1 changes the material. You can ensure that you're building the right way with all of these controls.

Building on the Xbox

The Xbox has different controls for building in the game. B is what opens the building menu, holding it will help you edit the structure and Y is what cycles through the building menu. RB will rotate for you while RT places the structure. If you want to change out the building material then LB. Right stick resets the building process.

When you open up the build mode, you can choose one of the structures and it will show up as a shadow-like holographic picture in front of the character before they begin building. You can then move the picture around to the right angle and area that you want it to be before you begin building it.

You have to keep in mind that when you're placing walls and other structures they do not automatically build, they do take a bit which can be frustrating if you're in the middle of a heated battle. The steel takes the longest to build up while wood takes the least time to go up. HP per second is the determining factor on how fast the materials build when you go to put them up.

Using skills and squad bonuses can help boost your HP and help you build in a much faster way which is beneficial when you play often and you want to be able to build that wall to protect yourself or to just take the health potion that might save your life. Many players work towards being the fastest builder in the game, with many of them already accomplishing this task.

ADVANCED BUILDING: CREATING COMPLEX STRUCTURES AND END-GAME FORTRESSES

Not only is this going to help you keep your distance and have some cover against the couple of other people you are against, but it is going to make sure that you have a place to go when you need to heal up in a pinch. You can create your own structure for this very need and it is known as an end-game fortress.

With a fortress at the end of the game, you might find yourself having a better chance surviving or at least coming out as one of the top 10 players, which is an accomplishment on its own in such a fast-paced game.

One of the biggest things you can do is to create a base structure that is a structure inside a structure, inside a structure. You want to make sure you have a place to see out of and have it up high so that you can easily reach out to the people below. You want to make sure that this is something that is done with ease when you are last. Keep in mind when you build this that it should be built towards the middle of the circle, as this is going to keep you ahead of everyone else and also keep you within boundaries when the storm starts moving in again.

"Pro-Building" is actually a thing and when it comes to creating the best structure, the most unique, the fastest and the most involved then you are considered a pro-builder. If you're just starting out building then this is definitely a title that you want to try to live up to when it comes to playing with advanced fortresses.

You can even edit a build that you have put together to create another opening or portion of the build. This is one of the best things about the build. You can make it into something else when the original thing has been used or might not be what you needed in the first place.

Making a sniper tower and then building out and over is thought of as an advanced or pro build. If you master this type of build then you can say that you're able to take on many of the other builds that are out there. A lot of people will play close to the ground and a good fortress will give you an advantage over all of them.

BUILDING TIPS!

If you're looking for a way to build and benefit from it as much as possible, then take a look at these tips to help you become a stronger Fortnite player that wins more frequently.

Many professional builders within the game give these tips to help others out when they're struggling with their building. You don't have to worry about not being able to make it to the end just because you're not the best builder. You don't have to build to win, but it can be a useful trick to have.

Keep these building tips in mind when playing!

Before you start building at the beginning of the game, you want to loot and go around taking off players. Towards the middle to the end of the game, go towards the middle of the game board and then build. This will hold up well over time and it will stay there for when you need to use it.

Make sure that the build that you put up is going to be inside the circle when the storm comes in. There is nothing worse then spending all of that time building a big structure only to find out that you cannot use it because you have to come in during the game.

If you can funnel the players around you that are left into a single entry point into the structure that you built then you want to make sure that you place all of the traps that you have looted. This is going to be a great way to get them and to take them out without actually having to have one-on-one combat with them which is always a risk best avoided.

It's important to realize that when you're looting items and scavenging for resources that it can become quite noisy. To keep yourself from attracting people nearby, you want to make sure that you try to keep it to a minimum. You can look around you to see if you have anyone near and then go at it. Use the time wisely and always forage in an area where you are safe to do so. Many people are taken out when they are getting materials for their build. You want to actually get to the build part before you're taken out too!

It is crucial that you learn how to build and fast when just starting out with the game. This is a skill that is going to help you out the most when it comes to playing and being able to make it out as the last one standing. Remember, the game is more about surviving, rather than killing so you need to be able to make it to the end when everyone else is running around taking others out!

Make sure to make the most of the build that you're doing to ensure that you're building the right tools for the occasion. If you're in the middle of a battle and you want to give yourself cover then this is the best thing to do. Quickly building has saved so many lives, so practice building as much as you can during fights... it can save your life!

When you're building within the game, you will quickly learn all that comes from being able to do so quickly and efficiently. You can learn new builds and new ways to get out of those sticky situations. When you become an accomplished builder your friends will come to rely on you to get them out of trouble too and you'll be a more helpful ally.

Take your time to learn the game. Everyone starts off as a noob, but once you get the hang of it, you may just become the world's next best player. This is a win-win situation since you're ready to go and now you have all of the best tips for building and coming out a victor in the end!

THE ISLAND

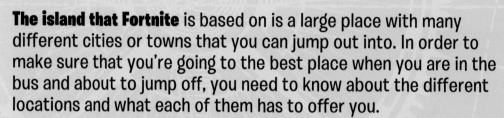

The island that Fortnite is based on is a large place with many different cities or towns that you can jump out into. In order to make sure that you're going to the best place when you are in the bus and about to jump off, you need to know about the different locations and what each of them has to offer you.

Each place has different structures, a different theme, different loot and some are much more popular than others. It is important to find out which area is offering you what you need before you decide to visit any of them. You need to find an area that feels the most comfortable for you and what you want. You don't want to land in an area that's going to make you lose prematurely, so take the time to get to know the Island and what different spots have to offer.

Keep reading and before you know it you'll be an expert on the different locations on the island.

WAILING WOODS

Wailing Woods is one of the best defensive locations on the map. You will find yourself hiding and barricading against oncoming enemies with a decided advantage. Located up on the northeast section of the map, it is definitely a place that is popular with the locals. Almost everyone drops in this area because it does have some pretty good loot.

Since this has a labyrinth around the middle of the bunker it is somewhat tricky to navigate the first couple of times but you will get the hang of it after a little bit.

Shipping containers to the right of the city provide the user with a way to scoop up a lot of loot and then hit the ground running. Some of the best items are found here, too. Expect to find some epic ranked items.

Another tip is to find a house that has been damaged. This little gem usually houses a chest and a bunch of weapons and ammo.

RETAIL ROW

Up on the middle right of the map you will find Retail Row. This is one of the most heavily populated areas at all times throughout the game. It has a bunch of loot, a bunch of hiding spots and a bunch of people that are willing to get in the middle of it all and fight to the death.

It is a retail section with multiple buildings next to each other and a parking lot to run through. The houses in the area also have chests and other loot up in the attics, so you can break through the tops of them to grab the stuff inside.

Balconies that are on the homes are great for ambushes for those that are leaving the retail complex and heading towards the houses. You can shoot them down from these areas before you're even noticed.

PLEASANT PARK

Found in the northwest of the map, you can expect to find some good loot in this area, as well. It is not as popular as Wailing Woods, but still offers the player a decent amount of opportunities. If you want something quick and easy, then this is the spot on the map for you.

Multiple little houses line the streets which makes it ideal for a quick and easy looting while moving around and remaining concealed.

If you want to scout the place for anyone nearby before you start looting, land on top of the gas station and you'll see everyone else in the area from your high vantage point.

LOOT LAKE

This is another spot that is well populated with multiple people and if you're not a seasoned player, then you might find that landing here is not the best option. You have to duck, dive, roll and twirl to get to the loot and then fight your way out of there.

There are a handful of structures that surround the lake which is where a lot of the loot can be found but many people enjoy going here since a lot of crates and chests are in the area and people drop off here since it is a localized point to the other areas on the map with some of the best looting points.

It is not advised that you cross the lake unless you know for certain that no one is in the area, as they can take you right out once they notice you.

MOISTY MIRE

Located in the southeast section of the map, this is one of the best places to go to get your feet wet in the yucky marsh. There's plenty of loot at this location and it's not as risky as some of the others on this list.

This is one of the less popular places to start and there is gear here, though it is not a lot. You can gear up pretty much uninterrupted and not have to worry about fighting off a bunch of people which is ideal for some.

Check out the prison located in the northwest corner of this area since this is where a lot of the loot can be found. You'll see loot in many of the cells and you'll gear up faster than you expect here. There is also a large loot tree that is in the area that holds around five chests per. You will want to check this out.

LONELY LODGE

If you're into the survival thing that comes with this game, then this is a great survival point for you to start in. You can be sure that you're getting the right hiding spots, but also that you don't have to drop and start fighting right away. This area allows you to collect some loot while developing a strategy to conquer all your enemies.

Check out the tower in this area since this is one of the best places to not only find some loot, but also to scope out opponents and perhaps take some of them out before they can even spot you. You can look over Retail Row from the top point and really let the other players have it if you happen to have a sniper.

There are a lot of hidden huts throughout the forest, which makes it ideal to check them out if you want some more loot to carry with you.

GREASY GROVE

Located on the southwest side of the map, you can expect to find a lot of residential houses in the area to loot. This is one of the more popular places to find the loot that you want and to see other players drop. If you want a decent amount of stuff and you want a good fight, then definitely stop in at Greasy Grove.

The restaurant and gas station are both ideal to spend some time in. You will find some chests and other tidbits that you will be able to use to your advantage.

In a corner of the restaurant, you can expect to find a three chest spawn that comes up and allows you to grab a bunch of loot. Again, be careful if you're new to all of this since this is one of the best areas to get into a firefight.

FLUSH FACTORY

If you want to be in the thick of it all then forget all of the spots that we say after this and before this. This is where everyone (and their brother) lands in the game. If you want to find a good fight, this is the place to go. Looting is tough and fast-paced, but there are tons of items to pick up here. Just keep moving and fighting because you'll find yourself dead otherwise.

You can end up here after the first rush goes by and still get a decent amount of stuff. It is not recommended to drop here before then, especially if you are a new player.

FATAL FIELDS

Mostly good for harvesting metal and other building materials, head here for crafting materials that you'll want later on in the game. There are some chests that spawn in the area but there are not too many, so it might be a good place to go after you have armed yourself somewhere else.

It is very similar to Anarchy Acres and it provides the same crop and field type of feel that you wouldn't be able to get from some of the other areas. Located in the south, you will find barns that have some items in them, but they can be collected for materials, as well.

One tidbit to keep in mind while playing here is that towards the bottom of the area, you will be able to find a truck that is labeled 'truck stop' and there are usually a few chests that are hidden within all of the rubble and mess.

ANARCHY ACRES

A lot like Fatal Fields, this is one of the best places if you're looking for barns, corn and scarecrows because they have a lot of all those. The house in the northern part of this area has a chest and a bunch of ammo according to some of the players.

You can check in the houses, barns and behind the haystacks to find things that will help you survive the rest of the game. It is not an extremely popular place to start, but it is definitely a place that you can go with a few others and not too many.

You might have a bit of a fight or you might have lucked out when someone else took all of the items first. It is also said that this is one of the best places to go because the spawn rate here is much higher than anywhere else in the game so you can continue to get awesome gear if you sit and wait it out, even after someone else has already been through here.

DUSTY DEPOT

This is a very popular area because this is where players are able to find top-tier loot hiding. You can be sure to find some good stuff but also some good fights from other players looking for the same types of loot. Make sure you're prepared and equipped to fight.

It is set towards the middle of the map and you can expect to find a lot of warehouses with many of chests and other things that you can loot within the area. To get the legendary gear you have to go in between the alleys outside the buildings. They're dark and dank but worth it.

Go to the southeast to a gated area where you'll still find plenty of loot while getting away from the clusters of people if the main area is overcrowded and you need to find something to fight with. The narrow spots are where a lot of the good loot is found.

HAUNTED HILLS

Just like the name suggests, this is a place where you can expect to find a cemetery with headstones and all. A mausoleum is even there for you to check out. Creepy, yes... worth it, absolutely! You will want to check this out whether you land in the area or are just passing through, sometimes you can find some pretty good loot hanging out here.

Located in the northwest corner, if you like playing here you'll have to do a bit of running to reach the location in many instances. It is not super popular as a landing spot, so you won't have to fight off a bunch of people.

Check the mausoleum and then snipe people from the top of it after you have been inside. You'll have an excellent vantage point to shoot from, giving you the advantage over everyone. Once you're finished looting, head over to Junk Junction to get a bit more loot before heading into the thick of battle.

Keep in mind that the storm eye moves and when it does, Haunted Hills is not going to be in the middle of it anymore. You'll have to move.

LUCKY LANDING

Located all the way to the south, this is a place where many players say is not very lucky, even though it says it in the name. It was a very populated area at one point since it is one of the newer spots on the map but it has definitely cooled down with some time, so it might be ideal for you to check out.

There are a number of smaller buildings that are around the area with loot, a large Cherry Blossom tree in the middle and a town hall that has pretty good chest spawns every now and again.

Since you're going to be by a decent place, you need to make sure to equip yourself as soon as possible, grab what you can and move on. Not only that, but as the storm moves in, this is going to be outside of the eye so you make sure to move with it, if not before it.

SALTY SPRINGS

This central spot makes looting and falling one of the most popular to go with. You can be sure that you're finding just about everything that you need, if you can make it through the houses that are lined along the streets and anything else you can think of.

Salty Springs is ideal for those that are willing to take the loot and fight for their lives, as this is a very popular area for many people to fall now. You can be close to the first circle storm that comes in, making it easier to get to the middle of the map.

Mostly good for loot, a lot of people will drop in, secure the area, take the stuff and then go somewhere else.

SHIFTY SHAFTS

This is definitely a cool place to check out, even if you're not looking to win the game. You have to be here at least once. It is a mining area that has underground tunnels which makes it cooler than cool.

You can find a lot of loot in the town, but also in the tunnels that are packed with loot. You'll probably even find a chest or two hiding down there. You want to keep up on your close proximity fighting as this is what happens a lot here since it is not a very big space to fight in.

It is also a great place to plan out ambushes since there are plenty of places to hide. Great if you're doing the ambushing, bad if you're the one being ambushed. Don't be caught off guard going through the tunnels and then running into someone else along the way.

SNOBBY SHORES

A residential area located on the west side of the map, you can expect to find lavish, posh houses that are gated. One of the coolest things is that there is a secret fall out or bomb shelter that is underground in the back of one of the large houses. The entrance to this is behind a dresser.

The area has a lot of people coming and going because of the rich surroundings. The area makes anyone want to loot it. You can expect to find a decent amount of loot, but also a decent amount of people so be prepared either way.

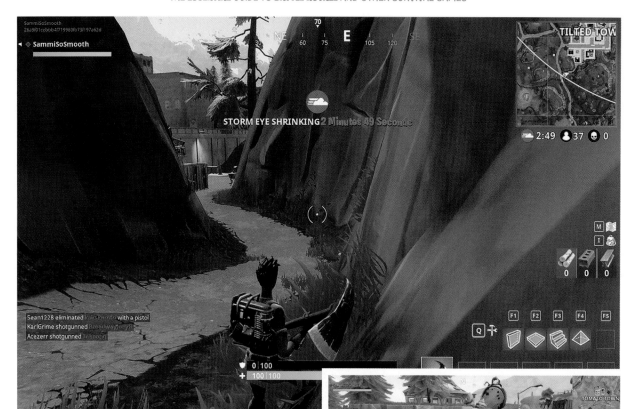

TILTED TOWERS

If you want a good view or close quarter combat then this is the place for you to go. You can climb up to the top of the towers and look down at everyone below. If you go to the underground portion of this area, there is usually a truck down there and it sometimes has loot in it for you to take advantage of. There's a chance you'll find a crate hidden here as well.

North of the park there is a brick on the ground that you can destroy. Make sure to do so and see if there is a crate that pops up for you to go through. This is a secret that a lot of players don't know. You just don't want to be out in the open when you're trying to get it to pop out, since you will be in plain sight.

The only downfall about this area is that you have to fight a lot of people. It is the most popular place to land because of all the hiding spots, the things to loot, the chests, secret areas and more. You might want to consider somewhere else if you don't want to fight for the items right away.

TOMATO TOWN

This smaller town has just a gas station and a restaurant but it is a town that does have some pretty cool items for you to loot. You want to check it out at some point. Just make sure to watch your surroundings so you don't get taken out right away. Be wary of snipers sitting on top of the gas station, because that's the highest point in this area.

There is a tunnel and sometimes you can find one or two hidden chests inside it. You will want to check out the cars, as well. Just make sure that if you mine them for metal that you don't activate their alarms.

It is definitely a low key area though, so you won't have too much competition when it comes to going through the many tunnels and areas around the town.

JUNK JUNCTION

A big ol' junkyard with a decent amount of loot, a lot of building supplies to gather and not a lot of people is what you will find within this area. Located at the northwest part of the map, you can expect to find a lot of crushed cars and some extra goodies waiting for you if you make this your starting point.

This is where you will find around 3 chests. Just make sure to keep in mind that there is a "llama" behind the junk that you have to watch out for when you make your way around the piles of cars. There is also no need for ramps in the area since you can jump from one car to the next right on the tops of them.

Just like the Haunted Hills, this is on the outside of the map towards the corner so you will want to keep your eyes on the storm since this is not going to be in the middle of the eye once it starts moving and you will have to move in with it.

BATTLE PASS

If you're looking at leveling up or unlocking some pretty epic gear and extras, a Battle Pass is a must-have addition to Fortnite. Without this neat little pass, you'll miss out on some of the best extras that the game has to offer. Not only that, but you are stuck with whatever they want to give you for a character.

Why not personalize the character that you have, grab some awesome emotes and even level up higher than you have ever leveled before?

The Battle Pass makes this all possible and you can unlock all of the extras that come with the game.

You just have to check it out for yourself and see if it is something that you'll get value from. There's a lot to enjoy about the pass, but some enjoy the features offered more than others. That's why it makes sense to examine all the benefits and decide for yourself!

The Battle Pass is definitely worth the cost if you like extras and want more from the game.

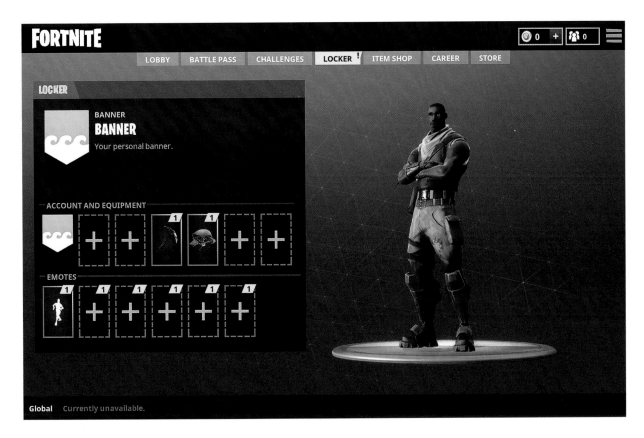

HOW AND WHERE TO HARVEST

If you're wondering why you should buy the pass, you're probably wondering if it's necessary to play the game. In short, you don't really need it if you don't want to level up further or get emotes, gestures, outfits, skins and more.

The more you play while on the Battle Pass, the more you're going to unlock through Fortnite itself. This is one of the best things about the pass and you can play at your own pace. Take your time going through all of the levels and tiers, and enjoy the experience that these add-ons offer you.

Weekly challenges are given to those with a pass, which adds something special to the game and makes it easier to progress. If you cannot play throughout the week, making it to your weekly challenge can be an ideal way to get a boost in a short period of time.

There are different seasons of Battle Passes, with each coming with different loot and extras that a person can unlock. Each season has its own set of tiers for you to rank up through as well, giving you serious goals to shoot for as you play the game.

Can't get the pass right at the beginning of a new season? That's okay, you can still get all of the season's rewards with the remaining time that you have! Each time you buy a Battle Pass you're going to get the full value out of it as long as you have time to play Fortnite and unlock everything.

Please keep in mind that a Battle Pass cannot be purchased with real currency. You have to use V-Bucks in order to purchase the Passes in every season.

A Battle Pass isn't required to have fun with Fortnite or to experience the survival content, but it's a nice extra that adds more to the game. The Battle Pass provides more challenges and gives the player more options when it comes to character customization so that you have the look that you want while playing the game. Without the pass, you can't take advantage of customization, challenges or many of the other perks because they simply aren't available.

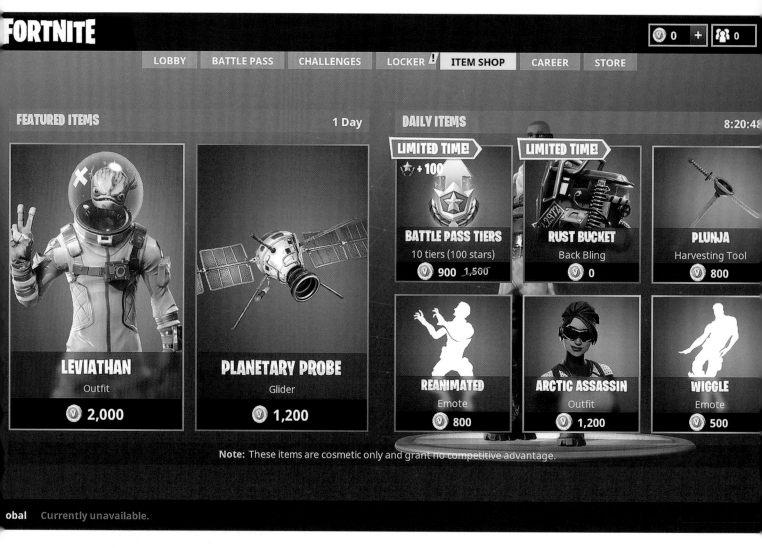

FORTNITE

LOBBY | BATTLE PASS | CHALLENGES | LOCKER ! | **ITEM SHOP** | CAREER | STORE

FEATURED ITEMS   1 Day

DAILY ITEMS   8:20:48

LIMITED TIME!

★ + 100

BATTLE PASS TIERS
10 tiers (100 stars)
900 ~~1,500~~

LIMITED TIME!

RUST BUCKET
Back Bling
0

PLUNJA
Harvesting Tool
800

LEVIATHAN
Outfit
2,000

PLANETARY PROBE
Glider
1,200

REANIMATED
Emote
800

ARCTIC ASSASSIN
Outfit
1,200

WIGGLE
Emote
500

Note: These items are cosmetic only and grant no competitive advantage.

obal   Currently unavailable.

WHAT COMES WITH THE BATTLE PASS?

Knowing what comes with the pass allows you to decide whether or not it's something that you want to go with. Of course, the items differ based on the current season, so it is important to keep this in mind. What is listed here might not be available when the next season comes out.

Battle Pass Season 3 offers a great deal more perks and content than what was offered in Season 2. For that reason, more players have been picking up the Pass and they're thrilled with all the content that they're getting. Many of the items offered in the pass were very difficult to obtain previously. In each of the categories of items being handed out, Season 3 had 2 to 4 more items than the previous seasons. Players are hoping this amount is going to be the new thing for all of the Seasons to come out from now on.

You can expect to get these items with the Seasons Battle Pass:

- Emotes
- Emoticons
- Gliders
- Pickaxes
- Outfits
- Skydiving Trails
- Back Blings
- Banners
- Loading Screens

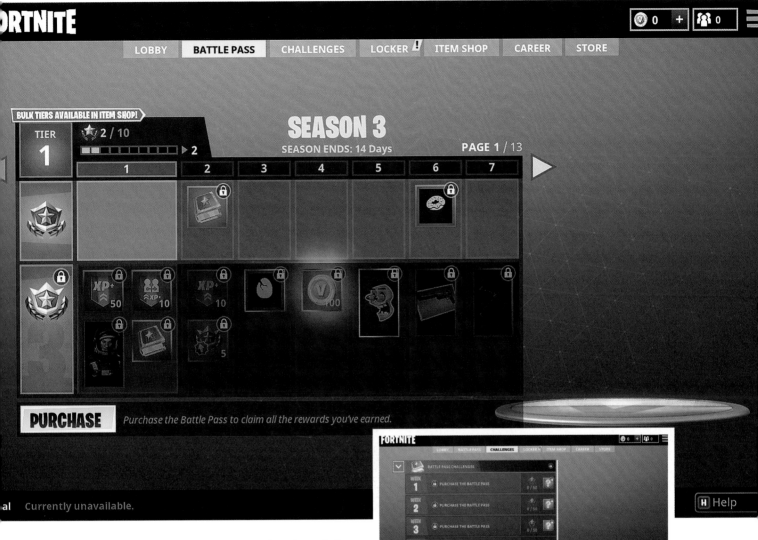

You get all of these items and get to level up through the use of the Battle Pass. If you do not get the pass, you still have the chance to acquire new items but you do not get as much and any exclusive or epic items are not handed out to non-pass holders currently. That's why many of the most serious players decide to pay for a Battle Pass.

When you sign up for the Battle Pass, you cannot see what is being given out entirely, but the makers do provide a glimpse at some of the epic items available for the season that you choose to sign up with. This brief preview makes it even easier to decide if the pass is worth the money for you or not, since you might see an item you just have to have.

PREMIUM CHALLENGES

Battle Pass holders have access to a feature called Premium Challenges as well when they play Fortnite. These challenges are offered on a daily, weekly and even monthly schedule so you can make the most of your available play time every single time that you log into Fortnite. Once the challenges are initiated you can play at your own pace, but just by being part of them you'll add extra chances to snag epic loot and benefit more than the non-pass holding players can.

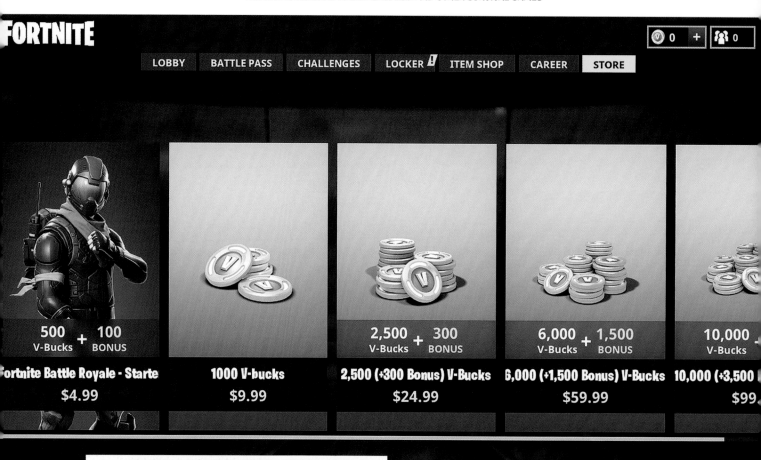

The creators of Fortnite do give default challenges for those that are regular players but don't have Battle Passes. This helps with leveling goals for everyone, but not to the same extent as pass holders. If you have a Battle Pass, then you're also offered a premium challenge on top of the standard one that all players get. If you do both of the challenges every day you'll get to the top tier of that season faster than standard players, and certainly quicker than the full three months a season lasts. That's a real privilege that will give you time to gather more exclusive items.

However, it's important to note that if you do not want to go through the challenges and beat them to get the stars, you can purchase the tiers using V-bucks. You cannot open access or challenges that are in seasons before or ahead of the season that everyone is currently on, but you can advance even faster than pass holders if you just buy the tiers yourself.

You also do not have to do the challenges to keep your Battle Pass access, everything can be done on your own time and when you have the chance to play the game and achieve the tiers.

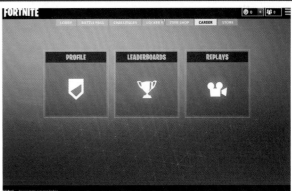

"SEASONS"

Every year seasons in Fortnite come and go, changing the experience for everyone. Each season comes with its own specific season pass, giving you access to the exclusive perks. Seasons usually run around 3 months at a time and give exclusive offers, items, gear and more for that particular time of the year. Sign up for a season pass during that time and you'll have access to content specific to that 3-month season.

There is also no limit to the amount of seasons that a person can sign up for. You can sign up for one season, grab those perks and then play as a normal player once again, or you can pay for passes for the next 4 or

5 seasons and grab all the exclusive content. Grabbing these passes is a great way to have access to all of the exclusive gear that you cannot pick up in the game or to move up through tiers when you want to level up and show your skills.

Note: Battle Pass rewards grant no competitive advantage.

You've already passed through the first few seasons of Fortnite, but there is still plenty of exclusive loot and perks for you to grab in future upcoming seasons. The creators of Fortnite are making more exclusive content than ever before, and it's a better time than ever to invest in a Battle Pass for the game.

Just remember that when you sign up for the current season that you are not able to get the items from the previous seasons that others have gotten. Those are

exclusively for those seasons and are usually not for sale at all. Many players sign up hoping that they can access older items that they like, only to be disappointed. The creators limit access to these items to make them more exclusive and exciting and to encourage players to participate in each season individually. It's a practice that keeps Fortnite fresh and exciting over time.

That doesn't mean you cannot work towards the items that are being handed out for the current and upcoming seasons! Don't miss out on the items that you want yet again. Get your hands on the current Battle Pass and pick up those exclusive items before they are gone for good! There are always new emotes, items and more coming out for the game, make sure you're around to get your hands on all of them.

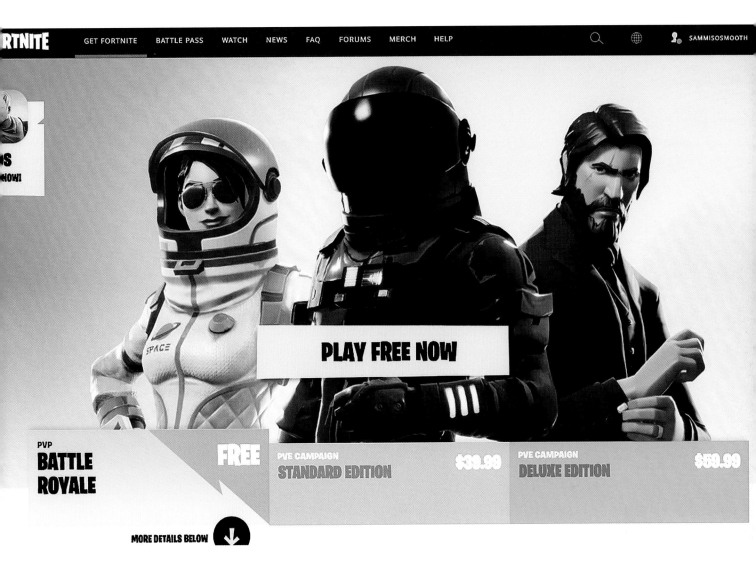

RTNITE   GET FORTNITE   BATTLE PASS   WATCH   NEWS   FAQ   FORUMS   MERCH   HELP   SAMMISOSMOOTH

PLAY FREE NOW

PVP
BATTLE ROYALE   **FREE**

PVE CAMPAIGN
STANDARD EDITION $39.99

PVE CAMPAIGN
DELUXE EDITION $59.99

MORE DETAILS BELOW

HOW TO BUY THE BATTLE PASS

If you're wondering how you can get on the game and purchase the Battle Pass, follow these easy instructions. Everyone can purchase a Battle Pass and you do not have to have any special type of account to do so. You just need a standard Fortnite account.

Here are the quick instructions for buying a Battle Pass when you're already an active player on Fortnite and you decide that you want in on those exclusive items.

1. Launch the Game
2. Select the 'Battle Royale' Option on the Screen
3. Go to the Battle Pass Tab on the Top of the Screen
4. Purchase the Battle Pass or the Battle Bundle

FREE	PVE CAMPAIGN STANDARD EDITION	$39.99	PVE CAMPAIGN DELUXE EDITION

PLAY FREE

BATTLE PASS
SEASON 3
PLAY BATTLE ROYALE. LEVEL UP. UNLOCK EPIC LOOT.

LEARN MORE

If you're purchasing later in the season then you will receive all of the items up to the point that you purchase at. Again, you have to purchase the pass with V-Bucks and not with actual currency but that's simple to do right through the store itself. Extra V-bucks can be purchased to buy other items that you may want to purchase in the store, but you only need a small amount of the V-bucks to purchase the pass or the bundle.

Battle Passes cost 950 V-Bucks which equals out to $10. That's an excellent value for the extra loot and tiers you get when you go through the extra Premium challenges. You'll be able to collect the stars and gather more loot with your pass, making Fortnite just a bit more exciting.

WHAT IS THE BATTLE BUNDLE?

While looking through the store you may have noticed that there's a Battle Pass and a Battle Bundle. They're different products and it's important to understand both before deciding which one you want.

When you buy the Battle Bundle you get the Battle Pass and all of the exclusive offers that come with it, as well as 25 extra tiers that you would not normally get with just the Pass. If you purchase this option then you get 40% off the purchase that you make. It is a cost savings and it also gets you a bit extra when it comes to cashing out on the pass and the items.

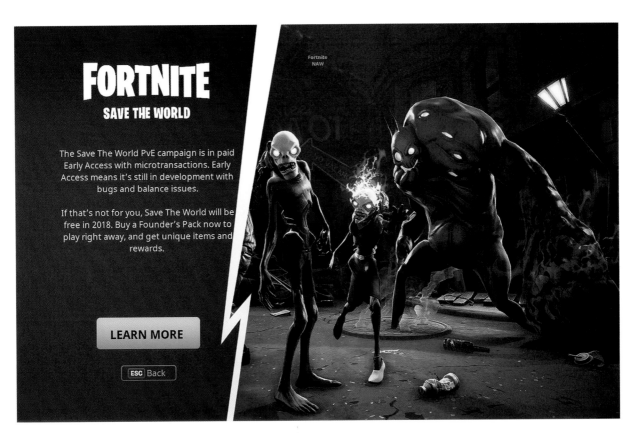

FORTNITE SAVE THE WORLD ACCESS

If you want to do something a bit different than the normal Battle Royale then the Battle Pass is an even better value because it gives you access to the unique game type called Save the World. This is a story mode that you can follow through that has the same weapons, gameplay and characters, but it offers a few different options that you can't get from the original Battle Royale version of the game.

This campaign version is a survival game that pits you against computer opponents rather than other players. You have to survive as you make your way through the world and find a way to beat the entire game. This is a world where you do not play with others though, which means that some might not like playing it.

It is completely up to you on what you choose to do but without the Battle Pass, you are not granted access into this world where you can go and kick some butt and take some names and follow the path to righteousness. Some players prefer this single player mode more than the traditional PvP mode that Battle Royale offers, but it's not for everyone.

There's a lot of talk around Battle Pass in Fortnite and players are constantly looking for ways to make the most of this unique perk. If you're interested in grabbing unique items, perks, challenges and leveling up through more tiers this is the only way to do so.

Anyone looking for extras from Fortnite will appreciate the Battle Pass. Both the Pass and the Bundle have exclusive offers that you cannot get while playing the game without them. Pick them up if you want to make sure that you have all of the offers, do the quests and check out what the game comes with entirely. If you just want to play Battle Royale with your friends and you don't care about all the extras the Battle Pass offers you, you can do that too.

Remember! The more you play with the Battle Pass, the more you unlock in the game! So get out there and play! ⭐

EMOTING

Dance moves you can see and acquire

There are so many things that you're able to do while you're a part of the Fortnite game force. Not only can you dance and do different gestures while in the game, but you can also use emoticons while playing for added fun and expressiveness.

This will put more personalization to your characters and might actually give you something worth partying about. You should probably try to make it to the end of the game before you start busting any celebratory moves, but until then, you can check out some of the cool moves that are out there to use to your advantage while playing Fortnite.

Keep in mind that not everyone has all of these moves and you may have to either earn them or purchase them. There is a bit more information on this below.

WHAT DANCE MOVES ARE THERE?

In the game, you can expect to find a series of dance moves. Some of these you will have to unlock through playing the game regularly since they aren't available to new players. Other ones you have to purchase through V-bucks.

If you're looking to bust a move while in the game, then make sure to look at some of these awesome moves that come with the characters when it comes to playing and also having a good time with the funky chicken.

FORTNITE
THE **ESSENTIAL** GUIDE TO **BATTLE ROYALE** AND OTHER SURVIVAL GAMES

Here are some of the awesome dance moves and emotes you can expect to find in the game but they are always adding new ones upon player requests!

- PURE SALT OR THE SALT BAE
- BEST MATES
- FRESH OR THE CARLTON
- THE BASIC OR TURK DANCE
- GANGNAM STYLE
- ELECTRO SHUFFLE
- FLOSS OR THE BACKPACK KID
- **1** THE WORM
- TAKE THE L
- BREAKIN'
- ROCKET RODEO
- REANIMATED
- DAB
- KISS KISS
- STEP IT UP
- FLAPPER
- MAKE IT RAIN
- FLIPPIN' SEXY
- **2** GUN SHOW
- FINGER GUNS
- FACE PALM
- SLOW CLAP
- JUBILATION
- BRUSH YOUR SHOULDER
- TRUE LOVE
- **3** THE ROBOT
- SALUTE

Here is a list of the emoticons, which are a sub-class of the emotes that you can do while in the game...

- RAGE
- PEACE
- HEART HANDS
- BULLSEYE
- LOL
- RIP
- CLAPPING
- MVP
- SALTY
- DANCE PARTY
- **4** FLEX
- BAITED
- EXCLAMATION
- STEALTHY
- WOW
- ON FIRE
- POTATO AIM
- IN LOVE
- **5** THIEF
- BOOMBOX
- ROCKET RIDE
- HEARTBROKEN
- GOOD GAME
- BUSH
- POSITIVITY
- AWWW
- A+
- THUMBS DOWN
- THUMBS UP
- 1HP
- HOT DAWG
- MAJESTIC
- 200 IQ PLAY
- HOARDER
- #1
- KABOOM
- **6** FLAMING RAGE

All of the emotes and emoticons are rated like the weapons and some of them are rare or uncommon while others are more common.

They're always adding more moves and this list is expected to grow with time. Hopefully it will add in some other moves that people are asking for!

HOW TO DANCE IN THE GAME

How you dance depends on where you're playing the game, as every console is different.

If you're on the Xbox or PS4 then you want to press the down button on the controller and it then gives you a bunch of options to choose from.

If you're playing on the PC, then keyboard's default button is B. This opens up the same menu of emote options.

You will have to move the analog stick and press X/A to make it work on the consoles or through clicking with a mouse if you're on a PC.

Some characters will only have one active emote that they can use when this option is chosen. If this is the case then the character will do this emote automatically when the button is clicked and there will not be an option wheel that would otherwise pop up.

CAN YOU GET MORE EMOTES?

If you only have one default emote then you usually want to know if you can unlock or purchase more through the game. This way, you can personalize the character a bit more and give them a bit of your own flavor.

There are ways that you can grab some new moves for that guy of yours...

The first option is to buy them. You will have to use V-bucks to do so in the item shop. They have many different ones to choose from inside the shop, so you will have plenty to spend those V-bucks on. Another way is by progressing with a Fortnite Battle Pass.

A Battle Pass can be purchased using V-bucks that you buy. This then helps the player progress through the game while you play and unlock the special outfits, gliders and emotes and other specialty items.

If you're thinking about the dance moves that are going to be the most effective for your character then be creative and have fun with it! Try out the dance moves before you purchase them, as well! You can see how they look!

Have fun dancing! ⭐

SQUAD MODE STRATEGY AND TIPS

Get the squad together! You're about to win this match!

The squad mode is a match that you can pair up with people and then take on the world around you. This means that you work together with these teammates to bring down the others that are working to destroy you and your teammates.

In the past you could hurt your teammates with friendly fire, but since the recent changes, this is no longer a problem so you don't have to worry about being too close to them when you shoot

your guns, this makes squad mode a bit more relaxed and also prevents griefing.

You're also able to see your teammates on the screen and know where they are on the map, as well as mark areas that you want them to go to while you're all playing together. This is a big thing since you will be able to easily and effectively communicate and bring the team to a central point through the use of the waypoints on the map. You can also check on their health and armor to find out how much each of your teammates has to work with.

Squad

Battle Royale

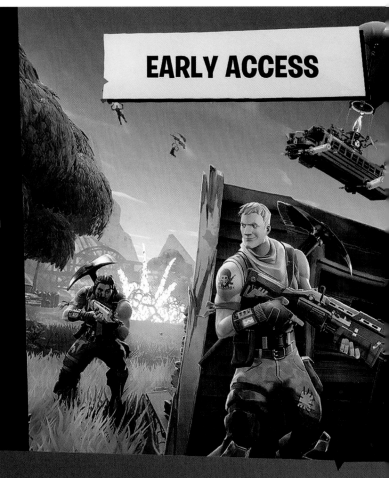

EARLY ACCESS

○ **Drop In & Load Up**

Pick a landing spot and search for gear.

○ **Stay Inside the Eye of the Storm**

Deadly clouds are closing in. The Eye will shrink as the storm intensifies.

○ **Last Squad Standing Wins**

There can be only one winner.

LOADING...

You can fall to your death. So... don't.

You also don't want to build massive bases since you're going to be pointed out, ambushed and all of your items are going to be taken and there ends the game for you. These are not specific to squad mode, either. They are ideal for any of the modes out there since they are good to know ahead of time before something actually does happen.

Playing solo is fun, but you are not going to have as much fun until you are playing in a squad and making sure to get more out of the game. You have never played Fortnite until you have some team members helping you out.

Strategize, strategize, strategize to win the game...

STRATEGY AND TIPS FOR SQUAD MODE

It is important that you know what to do and how to handle the situations that you might be faced with. Always make sure to keep these tips in mind when you're playing the game and you want to make sure that you get the right strategy out there that is going to make a difference.

First you want to make sure that you know where everyone is weak and where everyone is strong. You can give them the best jobs that match those skills. If someone is awesome at building things, you can make sure that this is done through the use of their skills, while someone that is excellent at sniper rifles having that under control and so on. Having a role for everyone is the best way to go when you have more than one person playing in the same game together. This is ideal for duos, too.

Always try to attack a full squad when you have good gear, good communication and when you are not attacking from the ground. Attacking a full squad from the ground is generally not an ideal way to take a whole team out. You will have to work something else out, whether you split your squad up or you just go from the high ranges overhead. You need to make sure that you're not losing the battle.

Use the time in the pre-game lobby to talk about tactics and who is doing what, when and how. This is especially helpful if you are just randomly matching up with people. You want to speak with them to find out what is happening and how it is happening. You will not be able to win the game if everyone is not on the same page. Use this pre-game time wisely.

Make sure to let everyone in your pack know what you picked up or left behind when looting. You want them to know what you grabbed and find out what others need so you can leave those behind if you find them for them to pick up. This will allow everyone to have a good amount of inventory stocked up for the entire game time and you won't have to worry about having ammo for someone else that they might need and run out of and so on.

Grab as many materials as you can while you are running through the areas. You need these to make sure to gain an advantage over your enemies through building. When you are running around, if you do not have them, you will not succeed in what you do. Make sure you and your teammates are always looting and gathering materials and weapons you need to win.

Do not, we repeat, do not run together in a giant clump. Not only does this call attention to your entire group but it is going to make it much easier for another group to take you out and harder for a teammate to hide and pick out the other team. Instead, stay apart from one another enough to hide individually and create more difficult targets for the enemies, but don't spread out so far that you can't help one another when trouble arrives.

Always look for and use opportunities to get up on your enemies. You want to make sure that you can sneak up on them and use the element of surprise in your favor. Be careful when approaching to avoid any traps that might be set, and always keep your eyes open for an ambush. You need to sneak up on your opponents the moment you realize you know where they are and they don't see you. Take them out sooner, rather than later. If you see them, you have the advantage and you are ready. Now is the right time to strike!

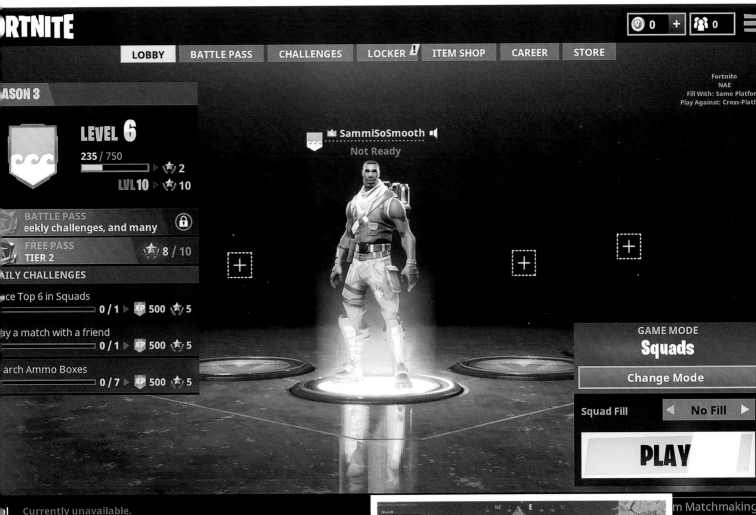

Always leave revives alone if you're playing with a group. You want to fight for yourself and make it safe before doing any revives. Make the mistake of trying to revive your buddy in the heat of battle and you can kiss your life goodbye. Fight for your life, and then try to revive your teammate after the action settles down.

Since it is a group effort, just because you took down one of the enemies does not mean that there are not more right around the corner. Make sure to keep your teammates close, and avoid fighting multiple opponents on your own whenever you can.

A quick tip to remember is that when an enemy falls and is removed from the match instantly, he has no more teammates left in the game. You don't have any more enemies from his squad to worry about and you can focus on looting in safety.

Some say that the most important rule to stick to is communication with the team. You don't want to do anything without conversing with the rest your teammates. You want to know that everyone is on the same page and that you're all going to use a strategy that you agree on. If you aren't talking you'll likely be thinking different things, and when you get into a battle you'll be at a disadvantage.

Be a team and be in the know. Make sure that you work with everyone on all fronts to get the job done. You all want to win, so you want to make sure that you're putting in your full effort to make the most of what is being provided. Be the one that makes it to the top of the game when you work together as a team and come out as one, too!

SQUAD FILL OPTION

"Squad Fill" is when you do not have your own friends to play with and you want the server to find random people to play with. This can be done when you do not have friends online or perhaps do not have friends to play with. You're easily able to find them when you choose this option.

Keep in mind that you're playing this game as a team. There is no I in team, as cheesy as that sounds, you need to make sure that you think about them and work with them as much as possible while running around. You might feel like you're going to do something great, but they need to know what you're doing as well. Being able to keep the lines of communication open, having a great tract record at what you do and making sure you're on top of it all can help you (and your team) win this round.

Work hard and work together and you can all come out victors! Just don't find yourself without a squad as this might make things harder! ⭐

MOBILE GAME TIPS

Yes! You can play Fortnite right from your phone! You just have to download the app and away you go! Of course this new feature won't work well for everyone that tries it, so test out Fortnite Mobile to see how the experience is for you personally. If you don't have a computer or game console mobile Fortnite could be an excellent opportunity for you to try this top-rated game out.

The mobile version is a bit different, since the screen is condensed but still just as cool. You will be able to bring the excitement of dropping from the bus and even taking out the other players right on your phone so you can enjoy the action from anywhere you want to play.

If you're ready to go and want to check out the mobile version, download it right to your phone and give it a go. If you're stuck, then come back here and check out some of this information that lets you know a bit about gameplay and some tips to get you started and help you survive just a bit longer.

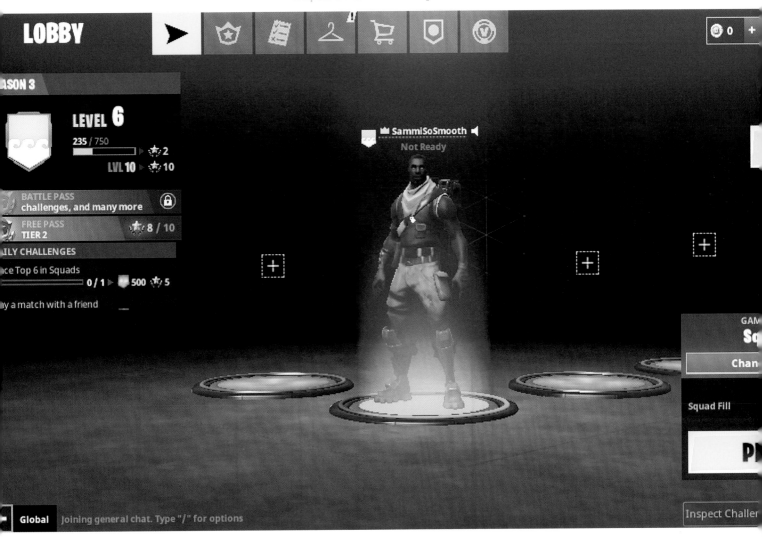

LOBBY

SEASON 3

LEVEL **6**

235 / 750 ▷ ⭐ 2

LVL 10 ▷ 🍃 10

BATTLE PASS
challenges, and many more 🔒

FREE PASS
TIER 2 ⭐ 8 / 10

DAILY CHALLENGES

Place Top 6 in Squads
0 / 1 ▷ 🛡 500 ⭐ 5

Play a match with a friend

👑 SammiSoSmooth 🔊
Not Ready

GAME
Sq
Chan

Squad Fill

PL

Global Joining general chat. Type "/" for options

Inspect Challer

PLAYING FORTNITE ON MOBILE

You have to download the app and then download the game to your phone once you open the app. Once you do that you can choose the mode that you want to play and set it up.

You do have to have an account or make an account in order to play. This is the same no matter where you choose to play. You have to adapt to the controls that are on the mobile device and think about how the Hunger Games works so that you can make it to the top at the end of the game as the last man standing, because who wouldn't want to be the last man standing?

Set up your game and get started!

GAMING TIPS FOR FORTNITE ON MOBILE

Keep these tips in mind when you want to play Fortnite on your mobile device. They are the same tips that you can use for each mobile experience, so get studying so you know how to succeed.

Be a master on the mobile version of this game and take down your opponents when you face off against them with help from these tips.

All of the controls that you need are right on the screen and you would just tap the controls that you want. Inventory opens up a larger inventory, combat mode helps you fight, the joystick helps you move yourself and so on. You can control everything just by tapping your screen and no external controller is required.

Make sure to tweak the touch control sensitivity that is on the sides of the screen. You don't want the controls to be too touchy, but you also do not want them to be under touchy. You need a comfortable medium that works with you.

Sound is still something that provides a great resource while on the mobile game. You will want to check out headphones or turn your volume up when you're playing. You will then be able to listen for other players nearby, listen for the chests that sing to you or anything else. Listening is a big part of this game, no matter what type of console you're playing on. The best part about the sounds is that when you hear them, there is a visual cue at the top of the screen that tells you where you can go and how far you are from the item.

There is an auto run feature that comes with the game on your phone, so you can make sure to use this when you're trying to cross a lot of ground in a little amount of time. This will keep your hands and eyes free while you're running to make sure that no one is going to take you out while you're running in an open field. Use this increase in freedom to help you make the best decisions that you can so you aren't sniped from a hidden opponent because you're out in the open when you shouldn't be.

Since the screen is only so big, you'll want to minimize your use of build mode until right at the moment you need to be building. Click the button and you'll have access to a range of build features at the bottom of your screen. Use these features to give you cover, create raised platforms and give you an edge in any firefight that you encounter.

Try to avoid long-range fights whenever possible on mobile, because sniping accurately is tough with smaller screens. Instead of trying to be able to hit people from far away, make sure to bring them closer. Hide more and rely on cover to keep fights at a comfortable distance.

There is an 'aim down sights' control on the screen during battle. When you use this, you can make fighting a lot easier. You might actually be able to win it if you use this, too. Since aiming and shooting is less intuitive than on the other consoles, you will need to make sure to use this function to help you out. Open your sights, lock onto your target and let them have it!

EXTRA TIPS TO STAY ALIVE WHILE PLAYING

These tips are great for not only the mobile version of the game, but also to play on any of the other consoles that are out there.

- Collect the ammo, materials, weapons and more as soon as you hit the ground and you need to make sure that you hit the ground fast. Be the first one down, grab what you need and see opponents before they see you.

- Build barricades against incoming fire from other players and make sure to outsmart them and hit them back, but harder.

- If someone has built stairs to go up or hide behind, just shoot the very bottom of them and the rest of the stairs are going to crumble below them, as well. This also means that if someone is at the top of those stairs that they built and you shoot them down, they are going to fall and instantly die when they hit the floor.

- A blue flare on the ground indicates that this is where loot is going to be dropped and you want to either wait it out, loot it and then remove it or leave it there to lure in other players while you hide and then take them out when they come running to it. The same thing goes for someone that has dropped and has yet to be looted.

- Don't save your shield potions, make sure to take them as soon as you find them since this is going to provide protection and not have you worry if you're going to be taken out or not. You are covered with the potions.

· The type of weapons you choose and pick up are going to make a difference on how far you get in the game. You need to have the best weapons and have them shown off when you are playing. To win and get to the top 10 players, you have to make sure to have the best colored ones. Epic and legendary are the best to look for.

· Make sure to keep in mind that it is every man for himself. If you see someone, take them out if you can safely do so. You don't want to leave anything for chance. If you let someone go during the game then you are going to have a hard time later on since you might have to face them.

Being able to play Fortnite wherever you are is one of the best and most exciting aspects of the mobile version. This version opens the game up to a new audience of players and it's a fun new way to enjoy your smartphone or tablet too!

When you are playing on the mobile version remember that everything is condensed. All you have to do is tap the button on the screen and it is going to open up and give you a wider screen with even more options. A lot of people started playing Fortnite this way and did not know how to play on the other platforms at all.

If you're switching to this version from one of the other platforms, remember to set it in a way that feels the most comfortable to you. It is quite different in the way that it feels so it is important that you make sure to have the right feel for the game out there.

Have fun and make it to the top 10 in the mobile version using our awesome tips!

TWITCH
TOP PLAYERS TO WATCH

If you are looking to watch Fortnite in action from one of the best gamers out there, then you want to take a look at some of these players. They are taking Fortnite by storm or just some of them are rather interesting to watch. You'll definitely pick up tips and tricks while watching these pros, and you'll find yourself applying ideas from these Twitch experts to help you win even more often in your own games.

Some of these players can even be found on YouTube, so that is something worth checking out when you're looking for players that are going to let you in on all of the action.

Looking for something great? Want to watch some of the best players out there? You'll find these leading players on Twitch for sure, and some have YouTube channels as well for you to check out.

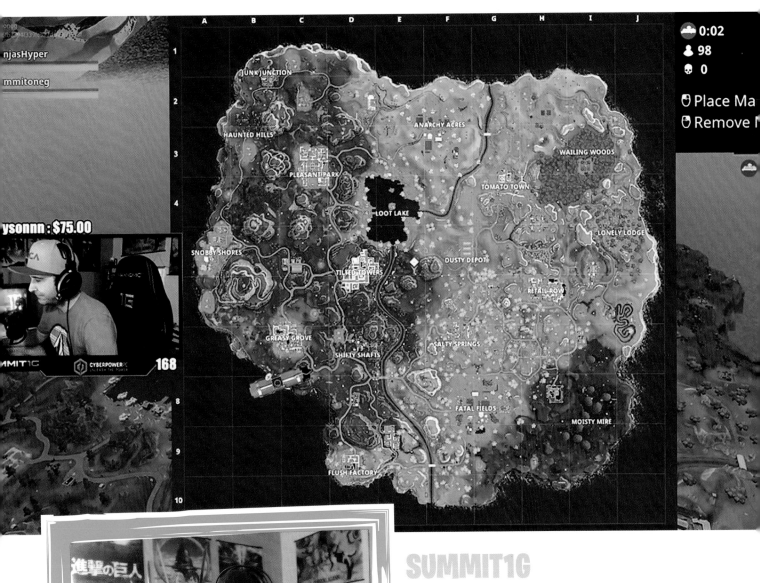

SUMMIT1G

One of the platform's biggest creators, he is definitely a nice guy and one you want to watch when you're on Twitch. If you're trying to get an idea of how the game works or feels, he has hundreds of videos up that you want to check into. Since he has been there to create them for all of this time, you can be sure that some hidden secrets are also there to check out. Who wouldn't want to?

MYTH

Known as one of the fastest builders to date for Fortnite, he is definitely a sight to see when you want to watch someone build fast and accurate. He gets involved in many high quality firefights and is an exciting player to watch for sure. He is also a top player with a record for consistently winning and regularly plays with the other Twitch members. He is definitely one you want to watch out for.

NINJA

One of the most entertaining Twitch and YouTubers that streams their games, no one contests that this is who they watch when they are watching someone play Fortnite. He is not only funny, but extremely entertaining, thought provoking, good at what he does and makes everyone want to play the game. He is all around the best channel to go to when you want to win some and learn some.

ALEXRAMI

He doesn't like to use a lot of emotes or even emotions when talking you through the play field, but he is able to teach you a thing or two. Very serious about his gaming, and it shows, he is one of the best players of the game as a whole. He has taken some time to get to know the game and be able to win at the end of it.

DAKOTAZ

One of the funniest streamers out there, he plays alongside HighDistortion a lot and provides the mix of entertainment that many are looking for. When they get together it is hilarious clip after hilarious clip. His name is also one of the more popular ones that you will hear when anyone mentions Twitch and Fortnite.

Twitch and YouTube are both filled with many gamers that are trying to get their videos seen. If you're looking to check out some awesome Fortnite players, then this is where you would be able to find them. You can check out what each of them has to offer. It might be something worth learning.

OTHER SURVIVAL GAMES

There are plenty of other survival games out there and some of them had their day while others might have squeaked by just barely. Being able to play some of these games will give you an idea of what each of them has to offer, and you might even enjoy some of them the same way that you enjoy Fortnite. In many you're racing the clock and fighting against a wide assortment of other people.

The type of games you have access to depends on what console or system you use, whether you are an avid mobile player or what you want to get from the game. Winning is a huge accomplishment, but most of these games have more goals than just winning and you'll quickly realize that there's a lot that goes into the typical survival game.

FORTNITE: SAVE THE WORLD

Fortnite: Save the World is another game much like Fortnite: Battle Royale except you get to play against the computer in a story mode instead of playing against a wide variety of other players out there. You'll be facing off against hordes of zombies and doing what you have to in order to survive.

If you want a different type of challenge and a little extra practice then this is where you want to go to get it. They have a bit of everything for the player and you don't have to worry about being online or matching with people since it is a story mode that you get to play on your own.

PLAYERUNKNOWN'S BATTLEGROUNDS

This is one of the biggest known battle royale type of games out there. With a lot of excitement about the game, it quickly went a bit downhill when Fortnite came out and blew everyone's socks off. It is still pretty fun and it has a lot of the same likeness that comes with Fortnite but they also have their differences too. Due to this, it is important to note that many of these battle games are a lot alike and they all have something unique to offer when it comes to their gameplay.

If you're looking for a game like Fortnite to play and you have not played PUBG yet, then you might want to give it a go. You will see why everyone mentions PUBG and why you might want to have an idea of what it is all about.

TROVE

Trove is Fortnite meets Minecraft. You are able to do the same type of battle for the world as you are in Fortnite but it is a sandbox type of game with building blocks that you would find in Minecraft. It is highly rated as one of the best games out there to play with multiple players and has been around for some time. It has gained a decent following in the past years.

Trove is also a game that you can expect to grow with some time and a little bit more attention put on it. It works in story elements and gives you a character that you can level up while unlocking better armor and more abilities for them. It's a lot of fun to play!

DARWIN PROJECT

This has the same feel and everything as Fortnite but it is not as large and it starts off a bit different. However, you are against 10 players and everyone fights to be the last one standing, which makes it exciting and a survival type game. It is post apocalyptic and the Ice Age is fast approaching so everyone is trying to stay alive as they fight their way to the resources.

This is said to be one of the closest games that resembles the Hunger Games, which makes it pretty fast paced and fun for those that want to fight to survive and come out on the top. It is only for Xbox One and PC though, so PS and mobile players are out of luck.

Darwin Project TM © 2018 Scavengers Studio.

RULES OF SURVIVAL

Though this game is not as good as some of the others mentioned, or as good as Fortnite, it definitely brings some promise with it that the others might not have had. It is a lot like Fortnite because you have to fight to survive against a bunch of other people. One of the biggest things about this game is that it is only available on mobile devices, though it is free to play. It has yet to reach its peak but it is definitely one to check out if you want to bring something to play with you on the go.

UNTURNED

One of the hottest games of the year when it came out, this is a game that is going to make you work hard to gather resources and build up what you need for long-term survival. This game is all about beating zombies, gathering materials, developing camps and becoming the best in the world. It's a single player game or an experience you can enjoy with other players. There are vehicles, ranged weapons and plenty of cool features to enjoy in this game.

The best part is that it is free to play and it is available on all computer platforms, so you can play with others through the internet.

INFESTATION: THE NEW Z

Another zombie survival game, this is one of those exciting games that has you running around and collecting the many different things that you need to survive. You take on zombies, as well as other players that might come your way and try to take your survival gear so that they can survive. Choose free play and PvP your heart away or follow the story that brings you to new places.

There are many games out there that you can check out. With many different stories and scenarios to follow, you can find a way to survive in all of the games that come your way. Never have to worry about not being able to find a game that you're good at and where you come out at the top. There has to be one out there for you!

CRAZY224

YOU PLACED #61

2:30

M Toggle Map   V View Match Stats   Report Player   Return

SammiSoSmooth

PUGZ1E

1

2:46 68

100 | 100
100 | 100

1

W   NW   N
GREASY GROVE
2:15 68

W   NW   N
SammiSoSmooth
WAITING FOR PLAYERS

0 | 100

0 | 100